"A complete and generous education, that which fits a man to perform justly, skilfully, and magnanimously all the offices of a citizen, both private and public, of peace and war." —MILTON.

CONTENTS

PART I

	PAGE
COLLEGE CALENDAR, 1918-1919	7
BOARD OF VISITORS	8
ACADEMIC BOARD	9
CLASS ROLLS	13
BATTALION ORGANIZATION	23
CITADEL CADET BAND	24

PART II

HISTORICAL SKETCH	27
LOCATION	31
BUILDINGS	31
ACADEMIC YEAR	32
MILITARY SESSION	32
FURLOUGHS	34
SYSTEM OF MANAGEMENT	34
RELIGIOUS SERVICES	35
LIBRARY AND READING ROOM	36
LITERARY SOCIETIES	36
Y. M. C. A.	37
ATHLETICS	38
ASSOCIATION OF GRADUATES	39
EXPENSES	40
LIST OF ARTICLES EACH CADET MUST BRING TO THE CITADEL WITH HIM	42

PART III.

REQUIREMENTS FOR ADMISSION	47
SYNOPSIS OF COURSES	52
DEPARTMENTS OF INSTRUCTION:	
Mathematics and Engineering	53
English	56
History	57
Chemistry	60

	PAGE
DEPARTMENT OF INSTRUCTION (*Continued*).	
Biology	63
Physics	63
Drawing	64
Modern Languages	68
Military Science	71
DEGREES	73
BENEFICIARY SCHOLARSHIPS	73
APPORTIONMENT OF BENEFICIARY CADETS	74

PART IV

ROLL OF GRADUATES	79
HONOR ROLL—GRADUATES IN MILITARY AND NAVAL SERVICES	107
EX-CADETS IN MILITARY SERVICE OF THE UNITED STATES	114
BLANK FOR HIGH SCHOOL CERTIFICATE	AT END
APPLICATION BLANK	

PART 1

COLLEGE CALENDAR, 1919-1920.

SESSION 1919-1920.

SEPTEMBER 20, 9 A. M.—Opening Day.

OCTOBER 21—Annual Review of the Corps by Board of Visitors.

NOVEMBER 27—Thanksgiving Day.

DECEMBER 18-23—First term examinations.

DECEMBER 23 to JANUARY 3—Christmas furlough.

JANUARY 3—Second term begins.

FEBRUARY 22—Washington's Birthday, holiday.

MARCH 29 to 31—Second term examinations.

APRIL 1—Third term begins.

JUNE 11 to 14—Final examinations.

ORGANIZATION.

BOARD OF VISITORS.

ORLANDO SHEPPARD, ESQ., Chairman_____Edgefield, S. C.
JNO. P. THOMAS, ESQ., _____Charleston, S. C.
JAS. H. HAMMOND, ESQ.,_____Columbia, S. C.
JAS. G. PADGETT, ESQ.,_____Walterboro, S. C.
REV. A. N. BRUNSON_____Columbia, S. C.

EX-OFFICIO.

HIS EXCELLENCY, R. A. COOPER, Governor, Columbia, S. C.

GEN. W. W. MOORE, Adjutant and Inspector General, Columbia, S. C.

HON. J. E. SWEARINGEN, State Superintendent of Education, Columbia, S. C.

J. HENRY JOHNSON, ESQ., Chairman Military Committee, Senate, Allendale, S. C.

J. L. MIMS, ESQ., Chairman Military Committee House of Representatives, Edgefield, S. C.

OFFICERS AND ACADEMIC BOARD.

COLONEL O. J. BOND, B. S., LL.D.,
Superintendent.

MILITARY STAFF.

LT.-COL. R. R. STOGSDALL, U. S. A., Retired.
Professor of Military Science and Tactics.
MAJOR JOHN W. MOORE,
Commandant of Cadets.
MAJOR E. M. TILLER,
Quartermaster.
FIRST LIEUTENANT C. L. HAIR,
Adjutant.
DR. R. S. CATHCART,
Surgeon.

ACADEMIC BOARD.

MAJOR ROBERT GIBBES THOMAS
Professor of Mathematics and Engineering.
Graduate, 1877; Instructor, 1878-1880; Professor of Mathematics, Engineering, and Physics, 1881-1882, Carolina Military Institute. Levelman, Western North Carolina Railroad and South Atlantic and Ohio Railroad, 1880-1881; Principal Assistant Engineer, Savannah and Tybee Railway, 1886; Resident Engineer, Georgia, Carolina and Northern Railway, 1888; Surveyor, 1883-1885; Assistant Engineer, 1885-1889, United States Engineers. Member of American Association for Advancement of Science; Society for Promotion of Engineering Education; Mathematical Association of America; and Allied Engineering Association of Charleston. Professor of Mathematics and Engineering, The Citadel, 1889. Chairman State Highway Commission, 1917.

MAJOR HUGH SWINTON MCGILLIVRAY, PH D., A.B
Professor of English.
A. B. College of Charleston, 1893; Student University of Munich, 1894-1895, and Ph.D. University of Gottingen, 1897; Head of Department of English, Charleston High School, 1898-1900; Professor of English, Converse College, 1909-1914; Professor of English, The Citadel, 1914—.

MAJOR JOHN WASHINGTON MOORE, B.S., M.A.
Professor of History and Political Science.
B.S. The Citadel, 1900; M.A. College of Charleston, 1913.

Commandant of Cadets and Instructor in Mathematics, Georgia Military Institute, Gainesville, Ga., 1900-1901; Commandant of Cadets and Master of Mathematics, Furman Fitting School, Greenville, S. C., 1901-1902; Instructor in Mathematics, University School, New Orleans, La., 1902-1903; Commandant of Cadets, University School, 1903-1904; Principal of High School, Greenwood, S. C., 1904-1906; Assistant Professor of English and History, The Citadel, 1906-1909; Professor of Political Science, The Citadel 1909-1913; Professor of History and Political Science, The Citadel, 1913; Major National Army, 1918.

MAJOR LOUIS KNOX, B.S., M.S.
Professor of Chemistry and Biology.
Student of St. Louis College of Pharmacy, 1896-1897; B.S., University of Texas, 1900, Chemist Texas Portland Cement Company 1900-1902; Professor of Chemistry Daniel Baker College, 1904-1907; Fellow in University of Chicago, 1907-1908; M.S. University of Chicago, 1908; Professor of Chemistry, The Citadel, 1908-1915; Professor of Chemistry and Biology, The Citadel, 1915—.

MAJOR LOUIS SHEPHERD LE TELLIER
Professor of Drawing and Assistant Professor of Military Engineering.
Graduate, 1903, and Post-Graduate, 1914, Engineering Course, Miller School; Student University of Virginia, 1904-1905; Teacher of Drawing and student in Engineering Classes, Miller School, 1905-1908; seven summers in engineering work. Member State Board of Architectural Examiners of South Carolina; associate-member American Society of Mechanical Engineers; Second Lieutenant, Infantry, U. S. Army, 1918. Professor of Drawing, The Citadel 1908—; Assistant Professor Military Engineering, The Citadel, 1917—.

MAJOR CLARENCE AUGUSTUS GRAESER, A.B., A.M.
Professor of French, German and Spanish.
A.B. College of Charleston, 1888; A.M. College of Charleston, 1896; Instructor in French and German, High School of Charleston, 1891-1896; Student at University of Gottingen, summer 1895; Superintendent Darlington Public Schools, 1896-1898; Instructor in French and German, High School of Charleston, 1898-1910; Student at University of Chicago, summer 1900; Student at University of Berlin, summer 1907; Student at University of Heidelberg, summer 1912; Instructor in French and German at Summer School, University of South Carolina, 1914, 1915, 1916, 1917; Professor of French and German, The Citadel, 1910—.

MAJOR FRANK ALEXANDER FERGUSON, A.B., A.M.
Professor of Physics.
A.B. University of Michigan, 1908. A.M. University of Michigan, 1914. Graduate Student, Johns-Hopkins Uni-

versity, 1914, 1916.; Instructor in Physics, High School, Bay City, Mich., 1914, 1916; Student Assistant, University of Michigan, 1907-1908; Professor of Physics, Mount Union College, Alliance, Ohio, 1908-1909; Professor of Physics, State Normal School, Oshkosh, Wis., 1910-1914; Graduate Assistant, Johns Hopkins University, Baltimore, Md., 1914-1916; Associate Professor of Physics, Carnegie Institute of Technology, Pittsburgh, Pa., 1916-1918; Professor of Physics, The Citadel, 1918—.

CAPT. CLIFTON LeCROY HAIR, B.S.
Assistant Professor of Mathematics.
B.S., The Citadel, 1909; Assistant Professor of Mathematics, The Citadel, 1909—.

CAPT. SMITH JOHNS WILLIAMS, A.B., A.M.
Assistant Professor of History.
Graduate, Normal Department, William and Mary College, 1903, and A.B. 1908; Graduate Student, Johns-Hopkins University, 1909-1910; Graduate Student, Columbia University Summer School, 1915, 1916, 1917, 1918; A.M., 1918, Columbia University; Grammar School Work, 1903-1907; Assistant Professor of English, William and Mary College, 1908-1909; Instructor, Virginia Summer Normal, 1909; Assistant Professor History, The Citadel, 1910—.

LIEUT. CARL FRANCIS MYERS, JR., B.S.
Assistant Professor of Mathematics.
B.S., The Citadel, 1914; Head of Commercial Department, Charleston High School, 1914-1918; Assistant Professor of Mathematics, The Citadel, 1918—.

LIEUT. ROSCOE HOWARD VINING, A.B., A.M.
Assistant Professor of English.
A.B. Boston University, 1916; A.M., Boston University, 1917; Department of Education, Penuelas, Porto Rico, 1907-1908; Public School Work, Massachusetts, 1908-1910; Student Boston University, 1910-1911; Principal, Union School, Tilton, New Hampshire, 1911-1914; Student, Bates College, 1914-1915; Student, Boston University, 1915-1917; Student at summer sessions, Hyannis Normal School, Massachusetts College of Agriculture, Dartmouth College, Boston University; Assistant in English, Boston University, College of Business Administration, 1917; Instructor in English and Modern Languages, New Hampshire College, 1917-1918; Assistant Professor of English, The Citadel, 1918—.

LIEUT. FLORIAN VURPILLOT.
Assistant Professor of French.
Born in France. Bachelier es Lettres, The Sorbonne, 1887; Laureate of University of Paris, 1889; Bachelor en Theologie protestante, University of Paris, 1891; Rector of the P. E. French Church of St. Sauveur, Philadelphia,

1895-1913; Assistant Professor of French Language and Literature, University of Pennsylvania, 1889-1913; Pastor of the Huguenot Church of Charleston, 1913—; Assistant Professor of French, The Citadel, 1918—.

LIEUT. ARTHUR MERRIAM CLARKE, B.A.
Assistant Professor of Physics and Chemistry.
B.A. Amherst College, 1917; Assistant Instructor of Physics, Phillips Exeter Academy, 1917-18; Analytical Chemist with Hercules Powder Company, 1918; Assistant Professor of Physics and Chemistry, The Citadel, 1919—.

MISS INEZ B. PARRY.
Librarian.

MRS. J. P. CHAPMAN
Matron of Mess Hall

MISS F. J. GASQUE
Matron of Hospital.

CARL METZ
Band Instructor.

CADETS, SESSION 1918-1919.

FIRST CLASS—SENIOR.

(*B, Biology; C, Chemistry; E, English; F, French; S, Spanish; M, Mathematics; P, Physics).

NAME	POSTOFFICE	*COURSES
ALEXANDER, C.	Chester, S. C.	E B S
BROWN, A. T.	Cross Hill, S. C.	M F
BUSH, L. E.	Ellenton, S. C.	M F
BUSH, M. L.	Greensboro, N. C.	M F
CANNON, T. C.	Honea Path, S. C.	E B S
COLEMAN, J. H.	Greenwood, S. C.	M S
COLEMAN, J. K.	Greenwood, S. C.	E B S
COTHRAN, F. E.	Greenwood, S. C.	M S
DILL, T. T.	Taylors, S. C.	M S
DUNBAR, T. E.	Ellentown, S. C.	M F
FORD, F. W.	Georgetown, S. C.	M S
FULLER, J. D.	Laurens, S. C.	C S
HAGAN, G. T.	Ora, S. C.	E B S
HANE, W. C.	St. Matthews, S. C.	C S
JEFFORDS, M. K.	Jonesville, S. C.	C B S
HART, J. B.	Orangeburg, S. C.	M S
JONES, H. C.	Walterboro, S. C.	M F
KEELS, J. W.	McColl, S. C.	M F
LAWSON, J. R.	Union, S. C.	C S
McGOWAN, J. C.	Cross Hill, S. C.	C B S
McMANUS, E. J.	Charleston, S. C.	M F
NICHOLSON, G. W.	Camden, S. C.	C B S
PEARLSTINE, M.	Charleston, S. C.	C B S
REYNOLDS, A. S.	Charleston, S. C.	C S
RIVERS, J. H.	Columbia, S. C.	C S
RUGHEIMER, E. W.	Charleston, S. C.	M S
SANDERS, J. H.	Sumter, S. C.	C B S
STILL, J. J.	Blackville, S. C.	M S
STREET, A. L.	Charleston, S. C.	M S
SURASKY, M.	Aiken, S. C.	M F
THOMPSON, F. A.	Batesburg, S. C.	C S
WANNAMAKER, W. W.	Orangeburg, S. C.	M S
WARLEY, S.	Charleston, S. C.	E B S

NAME	POSTOFFICE	COURSES
WILSON, J. W.	R. No. 1, Calhoun Falls, S. C.	M F
WITSELL, J. T.	Charleston, S. C.	M S
WOLFE, W. C.	Manning, S. C.	M S

SECOND CLASS—JUNIOR.

NAME	POSTOFFICE	*COURSES
ALLAN. G. H.	Summerville, S. C.	C F
ARTHUR, L.	Congaree, S. C.	M F
BAYNARD, R. S.	Landrum, S. C.	C S
BRADLEY, H. V.	Sumter, S. C.	M F
CARRINGTON, G. C.	Spartanburg, S. C.	M S
COOLEY, J. B.	Anderson, S. C.	
CRAWFORD, V. A.	Charleston, S. C.	C F
CROUCH, J. C.	Columbia, S. C.	
DOYLE, P. C.	Charleston, S. C.	C F
EVANS, WM., JR.	Bennettsville, S. C.	C F
GRIFFIN, J. E.	Walterboro, S. C.	M F
GROSS, M. E.	Holly Hill, S. C.	
HASELL, P. G.	Charleston, S. C.	M F
HAYNE, T. B.	Congaree, S. C.	M F
JACKSON, W. J.	Bowman, S. C.	M F
JAMES, W. E.	Darlington, S. C.	C F
KENDRICK, R. G.	Charlotte, N. C.	
LATIMER, T. C.	Chester, S. C.	M F
LINLEY, B. R.	Anderson, S. C.	M F
MAXWELL, W. J.	Florence, S. C.	C F
MIDDLETON, J. B.	Charleston, S. C.	C F
MILER, G. G.	Summerville, S. C.	M S
MOORE, C.	Waterloo, S. C.	M F
PADGETT, J. J.	Walterboro, S. C.	M F
RAINSFORD, J. C.	Edgefield, S. C.	M F
RILEY, A. W.	Allendale, S. C.	M F
SANDERS, S. M.	Charleston, S. C.	M F
SEYMOUR, R. E.	Greenwood, S. C.	M F
SMITH, E. B.	St. Louis Mo.	C F
SMITH, G. W.	Union, S. C.	C S
STEEL, G. H.	Evansville, Ind.	M S
STEWART, L. J.	Easley, S. C.	

NAME	POSTOFFICE	COURSES
THOMAS, J. P.	Columbia, S. C.	M F
WHALEY, W. E.	Edisto Island, S. C.	M F
WHITTEN, J. L.	Anderson, S. C.	M F
WILLIAMS, R. P.	Columbia, S. C.	M F

THIRD CLASS—SOPHOMORE.

NAME	POSTOFFICE
ALBERGOTTI, W. M.	Blacksburg, S. C.
ANTLEY, S. B.	St. Matthews, S. C.
ARTHUR, H. M.	Union, S. C.
BANNETT, A.	Allendale, S. C.
BECK, D. A.	Bradentown, Fla.
BETHEA, C. W.	Clio, S. C.
BRICE, W. O.	Winnsboro, S. C.
BRUNSON, R. L.	Florence, S. C.
BUYCK, W. F.	St. Matthews, S. C.
CARTRETTE, J. P.	R. No. 1, Allen, S. C.
CHENEY, B. B.	Lumber City, Ga.
COLEMAN, I. M.	R. No. 2, Pacolet, S. C.
COOPER, L. G.	Wilmington, N. C.
CUDWORTH, L. M.	Atlanticville, S. C.
DIAL, G. L.	Columbia, S. C.
DOTTERER, W. A., JR.	Charleston, S. C.
DUNN, C. A.	Camden, S. C.
FISHBURNE, T. R.	Blackville, S. C.
FRANKLIN, M. A.	Lima, O.
GARY, T. W.	Bartow, Fla.
GOODING, W. J., JR.	Hampton, S. C.
GIBSON, W. G.	R. No. 2, Gibson, N. C.
HAMPTON, A. G.	Lykesland, S. C.
HARTER, W. H.	Fairfax, S. C.
HARVEY, V.	Pinopolis, S. C.
HAYES, C. P.	Richmond, Va.
HEINSOHN, A. F.	Charleston, S. C.
HILL, B.	Cross Hill, S. C.
HUTCHINS, C. H.	Spartanburg, S. C.
INMAN, W. E.	Florence, S. C.
JAEGER, H. J.	Florence, S. C.
JAMES, H. M.	Summerton, S. C.

NAME	POSTOFFICE
JARVIS, R. B.	Charleston, S. C.
KAMINER, H. G.	Gadsden, S. C.
KELLY, J. O.	Manning, S. C.
LANGLEY, W. C.	Columbia, S. C.
LEWIS, J. S.	York, S. C.
LUCAS, W. C.	Charleston, S. C.
MCLEAN, P. J., JR.	Aiken, S. C.
MARSHALL, W. F.	York, S. C.
MARTIN, O. M.	Mullins, S. C.
MOOD, O. C.	Summerton, S. C.
MORGAN, E. R.	North Augusta, S. C.
MURDAUGH, H. V.	Columbia, S. C.
PALMER, B. M.	Cartersville, S. C.
PERRY, E. C.	Johnsonville, S. C.
PLATT, J. L., JR.	Mullins, S. C.
POLLOCK, E. A.	Augusta, Ga.
POULNOT, C. F.	Charleston, S. C.
REEVES, G. F.	Cottageville, S. C.
RICHARDS, G. P.	Charleston, S. C.
RILEY, G. O.	Barnwell, S. C.
ROBERTS, R. C.	Ehrhardt, S. C.
ROSS, T. W.	Florence, S. C.
RUFF, T. K.	Columbia, S. C.
SANDERS, P. W., JR.	Charleston, S. C.
SANDERS, S., JR.	Sumter, S. C.
SCOVILL, J. P.	Orangeburg, S. C.
SHEALY, M. Q.	Perry, S. C.
SKINNER, J. B.	Pinopolis, S. C.
SIMONS, B. W.	Charleston, S. C.
SINGLETON, B. N.	Westminster, S. C.
SMITH, W. B.	Bishopville, S. C.
SMITH, W. M.	Ridgeland, S. C.
TIEDEMANN, J. A.	Charleston, S. C.
WATSON, W. W.	Ridge Spring, S. C.
WATTS, B. S.	Cross Hill, S. C.
WHALEY, J. D.	Edisto Island, S. C.
WHITAKER, L. C., JR.	Charleston, S. C.
WILKINS, A. C.	Kingstree, S. C.
WILLIAMS, W. EARLE	Hartsville, S. C.
WILLIAMS, W. H.	Allendale, S. C.

NAME	POSTOFFICE
WILLIAMSON, T. W.	Florence, S. C.
WISE, G. C.	Orangeburg, S. C.
WITSELL, C. A.	Walterboro, S. C.
WORKMAN, P.	Rock Hill, S. C.
WULBERN, F. M.	Charleston, S. C.
YARBOROUGH, T. W.	Mullins, S. C.

FOURTH CLASS—FRESHMAN.

NAME	UNIT	SCHOOL	POSTOFFICE
ADAMS, E.	12	Ninety Six, H. S.	Ninety Six, S. C.
ALEXANDER, D. B.	13	Greenwood, H. S.	Greenwood, S. C.
ALLEN, J. R.	12	Washington, H. S. & S. M. A.	Washington, O.
ALLISON, C. F.	15½	Chester H. S.	Chester, W. Va.
ARTHUR, J. H.	14	Union H. S.	Union, S. C.
ASBILL, D. S.	14	Ridge Spring H.S.,	Ridge Spring, S.C.
AULL, L. B.	16.3	Ninety Six H. S.	Dyson, S. C.
AUSTIN, B. L.	14.8	Wagener H. S.	Wagener, S. C.
BAILES, T. E.	12	Bingham M. S.	Anderson, S. C.
BAILEY, G. M.	12	Bethel Col. Acad.	R 1, Olmsted, Ky.
BAILEY, J. N.	17½	Greenwood H. S.	Greenwood, S. C.
BARKER, H. B.	13.6	Allendale H. S.	Allendale, S. C.
BARRON, R. E.	14	Winthrop Tr. Sch.	Rock Hill, S. C.
BETHEA, P. O.	15	Dillon H. S.	Dillon, S. C.
BIEL, C. H.	12	Charleston H. S.	Charleston, S. C.
BIGGS, F. D.	17.9	Latta H. S.	Latta, S. C.
BLACK, E. W.	12.1	Walterboro H S.	Walterboro, S. C.
BLOOD, W. E.	13	Mulberry H. S.	Mulberry, Fla.
BONEY, S. M.	14	Columbia H. S.	Columbia, S. C.
BOWMAN, R. A.	12	Orangeburg H. S.	Orangeburg, S. C.
BROWN, E. E.	15.5	St. John's Acad.	Darlington, S. C.
BROWN, J. L.	16	Charleston, H. S.	Charleston, S. C.
BROADWAY, O. H.	12	Pinewood H. S.	Pinewood, S. C.
BROKENSHIRE, R. H.	12.1	Pawtucket H. S.	Charleston, S. C.
BRUNSON, J. E.	16.3	Ninety Six H. S.	Ninety Six, S. C.
BRYAN, J. L.	12	Columbia H. S.	Columbia, S. C.
CARD, H.	17.1	Augusta H. S.	Augusta, Ga.
CHAPLIN, S.	12.5	Charleston H. S.	Charleston, S. C.
CHASE, J. A.	12	Florence H. S.	Florence, S. C.
CLARK, W. A.	14.5	Grensboro H. S. & P. M. A.	Charleston, S. C.

NAME	UNIT	SCHOOL	POSTOFFICE
CLYMER, H. L.	20.5	Phoenixville H. S.	Phoenixville, Pa.
COGGESHALL, W. D.	16	St. John's Sch.	Darlington, S. C.
COHEN, I.	14	Charleston H. S.	Charleston S. C.
COLEMAN, J. W.	14	Greenwood H. S.	Greenwood, S. C.
COLONNA, R. P.	16	Accomac, Va. H. S.	Tasley, Va.
COPE, J. B.	12	Beaufort H. S.	Port Royal, S. C.
CROUCH, H. W.	14	Charleston H. S.	Charleston, S. C.
CULLER, T. R.	12.4	Cameron H. S.	Cameron, S. C.
DANTZLER, J. E.	12.8	Cameron H. S.	Cameron, S. C.
DANTZLER, M. O.	13	Orangeburg H. S.	Orangeburg, S. C.
DAVIS, W. I.		Wake Forest Coll.	Wilmington, N. C.
DAVIS, W. P.	13	Orangeburg H. S.	Orangeburg, S.C.
DELORME, C. C.	13	Darlington, H. S.	Dovesville, S. C.
DILTZ, L. E.	12	Spartanburg H. S.,	Spartanburg, S. C.
DOUGLAS, M. D.	12	Mt. Zion Inst.	Winnsboro, S. C.
EADDY, W. J.	12.5	Hemingway H. S.	Hemingway S.C.
EDDY, A. J.	18.5	Ninety Six H. S.	Ninety Six, S. C.
EDGERTON, J. B.	12	Florence H. S.	Florence, S. C.
EDWARDS, F. B.	12.5	Batesburg H. S.	Batesburg, S. C.
EDWARDS, M. W.	17.5	St. John's Sch.	Darlington, S. C.
EDWARDS, W. G.	12.5	Batesburg H. S.	Batesburg, S. C.
EISENMANN, L. R.	14	Charleston H. S.	Charleston, S. C.
ENDE, C. F.	19¾	Greenville H. S.	Greenville, Tex.
ERCKMANN, H. L.	12.8	P. M. A.	Charleston, S. C.
EVARTS, N. F.	15.5	Chester, H. S.	Chester, W. Va.
FAIREY, W. F.	17	Orangeburg H. S.	Orangeburg, S. C.
FANT, F. H.	13	Savannah H. S.	Savannah, Ga.
FELDER, E. W.	12	B. M. I.	R. F. D., St. George, S. C.
FIELD, C. E.	17.5	Bamberg H. S.	Bamberg, S. C.
FOGARTY, C. A.	13	Florence H. S.	Florence, S.C.
FORD, W. L.	14	Winyah H. S.	Georgetown, S. C.
FORTSON, S. D.	18	Richmond Acad.	Augusta, Ga.
FOSTER, E. E.	12	Charleston H. S.	Charleston, S. C.
FOX, H. B.	13	I. N. M. T. Sch.	New Orleans, La.
FREDERICK, F. J.	13	Marshallville H. S.,	Marshallville, Ga.
FREE, H. L.	13	Bamberg H. S.	Bamberg, S. C.
FROST, J. D.	13.5	Columbia H. S.	Columbia, S. C.
GALL, J. C.	16	Leesville H. S.	Johnston, S. C.
GALL, V. R.	16	Leesville H. S.	Johnston, S. C.
GASQUE, Q. D.	13.7	Laurens H. S.	Laurens, S. C.

NAME	UNIT	SCHOOL	POSTOFFICE
GEE, J. W.	16	Florence H. S.	Florence, S.C.
GLENN, E. B.	12	B. M. I. exam.	R 7, York, S. C.
GRAMLING, J. L.		Wofford Coll.	R 6, Orangeburg, S. C.
GREEN, A. H.	14	Sumter H. S.	Sumter, S.C.
GREEN, H. F.	13.5	Wilmington H. S.	Wilmington, S. C.
HAMES, S. T.	18	Union H. S.	Union, S. C.
HAMILTON, H. P.	11.7	Chester H. S.	Chester, S. C.
HANAHAN, J. R.	16	Charleston H. S.	Charleston, S. C.
HARDEMANN, W. L.	12	Newberry H. S. & Prep. Dept. Newberry College	Newberry, S. C.
HARLEY, H. C.	15.5	Orangeburg H. S.	Orangeburg, S. C.
HARRALL, M. M.	15.8	Wofford F. S.	Darlington, S. C.
HARRIS, C. H.	14	Baylor Sch. & Marion Inst.	Chattanooga, Tenn.
HARVIN, A. C.	13	Manning H. S.	Manning, S. C.
HECKLE, H. N.	12.5	Anderson H. S.	Anderson, S. C.
HESTER, W. H.	12	Albany H. S.	Albany, Ga
HORTON, T. E.	18	Columbia H. S.	Columbia, S. C.
HOWELL, T. M.	12	Walterboro H. S. exam.	Walterboro, S. C.
HUGHES, F. W.	13	Central H. S., Calgary, Can.	Walterboro, S. C.
HUGUENIN, T. A.	15.5	S. M. A.	Summerville, S. C.
HUMME, W. F.	12	Charleston H. S.	Charleston, S. C.
JACOBS, E. M.	12	Walterboro H. S., exam.,	Walterboro, S. C.
JEFFORDS, D. E.	12	Florence H. S.	Florence, S. C.
JEFFORDS, G. E.	17.5	Florence H. S.	Florence, S. C.
JETER, R. C.	16.5	Union H. S.	Whitmire, S. C.
JOHNSON, E. L.	13.5	Dothan N. C. H. S.	Loris, S C.
JONES, E. M.	12	Walterboro H. S., exam.	Walterboro, S. C.
JONES, G. O.	15.5	McCallie Sch.	Chattanooga, Tenn.
JONES, W. B.	12.7	Chester H. S.	Chester, S. C.
KIENZLE, T. C.	16	Louisville H. S.	Louisville, Ky.
KIRKLAND, B. B.	14	B. M. I., Hyatt Park Sch.	Columbia, S. C.
KLENKE, C. F.	14	Charleston H. S.	Charleston, S. C.
KNOX, J. H.	12.2	Williamston H. S.,	Williamston, S. C.

NAME	UNIT	SCHOOL	POSTOFFICE
KOLLOCK, O. H.	16.5	St. John's Sch.	Darlington, S. C.
LAKE, W. J.	12¾	Laurens H. S.	Laurens, S. C.
LANGFORD, P. L.	14	B. M. I., Prosperity H. S.	Prosperity, S. C.
LEE, R. E.	13	Florence H. S.	Florence, S. C.
LEE, W. S.	14	Charlotte H. S.	Charlotte, N. C.
LELAND, W. A.	12.5	Johnson City, Tenn. H. S.	Mt. Pleasant, S. C.
LEVI, M. H.	12.8	Manning H. S.	Manning, S. C.
LINDSAY, E. L.	12	Florence H. S.	Florence, S. C.
LITTLE, G. T.	12	Camden, H. S., et al.	Camden, S. C.
McALISTER, F. R.	15	Iva H. S.	Iva, S. C.
McCALL, W. S.	16	Florence H. S.	Florence, S. C.
McCORKLE, W. H.	13.5	York H. S.	York, S. C.
McGEE, T. Q.	13	Hastoc Sch.	Spartanburg, S. C.
McINTOSH, A. S.		University S. C.	Columbia, S. C.
McKELLAR, B. F.	12	B. M. I.	Greenwood, S. C.
McMASTER, J. H.	12.5	Mt. Zion Inst.	Winnsboro, S. C.
MACDONALD, J. C.	13	Columbia H. S.	Columbia, S. C.
MARTIN, J. J.	13	Anderson H. S.	Belton, S. C.
MASSEY, H. B.	17	Sandersville H. S.	Sandersville, Ga.
MAYFIELD, T. M.	13.8	Greer H. S.	Greer, S. C.
METZ, C. H.	12	Charleston H. S.	Charleston, S. C.
MILLER, W. Y.	12	MacGuire School, exam.	Richmond, Va.
MITCHELL, A. H.	12	Benton H. S.	West End, Ill
MOMIER, R. H.	15	Charleston H. S., B. M. I.	Charleston, S. C.
MOORE, D. W.	12	Greenville H. S.	Greenville, S. C.
MOORE, O. R.	12	Winthrop Tr. Sch.	Rock Hill, S. C.
MORGAN, L. T.	16	Suwanee H. S.	Brunswick, Ga.
MUCKENFUSS, H. B.	14	Clemson Coll.	Charleston, S. C.
MUNNERLYN, B.	14	Winyah H. S.	Georgetown, S. C.
NIXON, J. G.	12.5	Hertford H. S.	Hertford, N. C.
OSWALD, R. D.		Clemson Coll. R 1,	Charleston, S. C.
PEARCE, C. C.	14	Columbia H. S.	Columbia, S. C.
PEARCY, W. C.	12	Walterboro H. S., exam.	Walterboro, S. C.
PENDERGRASS, R. C.	13.5	Monroe H. S.	Monroe, Ga.
PHELPS, A. C.	15	Sumter H. S.	Sumter, S. C.

NAME	UNIT	SCHOOL	POSTOFFICE
POLIAKOFF, M. T.	12	Lancaster H. S.	Lancaster, S. C.
POLLOCK, D. M.	13.5	Monroe H. S.	Monroe, Ga.
PORTER, E. T.	14	Winyah H. S.	Georgetown, S. C.
POULNOT, L. S.	12	Charleston H. S.	Charleston, S. C.
POWERS, D. S.	17	Owensboro H. S.	Hawesville, Ky.
PRENTISS, C. B.	12	Charleston H. S., P. M. A.	Charleston, S. C.
RAUCH, C. H.	12	Chicago, Austin H. S.	Chicago, Ill.
REDFEARN, E.	12	B. M. I.	Pageland, S. C.
RICHARDSON, J. M.	12	Summerton H. S., exam.	Summerton, S. C.
RICHARDSON, J. P.	12	Summerton H. S., exam.	Summerton, S. C.
ROBINSON, G. S.	14	Carlisle Sch.	Charleston, S. C.
ROGERS, J. E	12.5	Charleston H. S.	Charleston, S. C.
ROGERS, L. B.	12	Wilmington H. S., North State	Wilmington, N. C.
ROPER, S. M.	12	W. N. & I., Hyatt Park; Columbia H. S.,	Columbia, S. C.
ROWLAND, W. W.	14	Sumter H. S.	Sumter, S. C.
SALLY, N. R.		Clemson Coll.	Salley, S. C.
SAXON, K.	12.5	Topeka H. S.	Topeka, Kans.
SEABROOK, E. M.		Clemson Coll., R 1, Charleston, S. C.	
SEABROOK, T. H.		Clemson Coll., R 1, Charleston, S. C.	
SESSIONS, C.	12	Horry Indus. Inst.	Conway, S. C.
SHERMAN, C. C.	16	Lyons Township H. S., LaGrange, Ill.	
SHULER, M. P.	16.5	Orangeburg H. S.	Rowesville, S. C.
SIMMONS, J. W.	12	Charleston H. S., P. M. A.	Charleston, S. C.
SMITH, A. T.		Furman Univ.	Greenville, S. C.
SPIGENER, P. S.	12	Greenwood, H. S. B. M. I., exam	Ridge Spring, S. C.
STANLEY, G. W. H.	16	Thomasville H. S., Thomasville, S. C.	
STEADMAN, K. C.	16.5	Chester H.S., New Cumberland, W.Va.	
STEPHENS, G.	13.5	Myers Sch.	Birmingham, Ala.
TAMSBERG, C. T.	18	Charleston H. S., B. M. I.	Charleston, S. C.
TAYLOR, E. B.	12	Florence H. S.	Florence, S. C.
THOMPSON, E. T.	17¼	Dillon H. S.	Dillon, S. C.
THOMSON, J. W.	16	Anderson H. S.	Anderson, S. C.

NAME	UNIT	SCHOOL	POSTOFFICE
TIFT, R. A.	15	Albany H. S.	Albany, Ga.
TIMMONS, H. A.	18	Florence H. S.	Florence, S. C.
TOLBERT, E. L.	12	B. M. I.	Greenwood, S. C.
TOLLESON, W. W.	13.5	Gaffney H. S.	Gaffney, S. C.
TREVATHAN, C. H.	14	Rocky Mount H. S.,	Rocky Mount, N. C.
TUCKER, WM.	12	Webb Sch.	Ripley, Tenn.
TURNER, J. Y.	15.5	W. F. S.	Winnsboro, S. C.
USSERY, E. B.		Furman Univ.	Elko, S. C.
WAGENER, W. Y.	13	Summerville H. S.,	Summerville, S. C.
WAGONER, A. B.	20.5	Staunton M. A.	Spring City, Pa.
WALLACE, J. W.	18	Florence H. S.	Mars Bluff, S. C.
WARDLAW, J. G.	13.5	York H. S.	York, S. C.
WARREN, C. B.			Williams, S. C.
WATSON, T. H.	14	Whitmire H. S.	Whitmire, S. C.
WEBSTER, W. M.	13.5	Hastoc Sch.	Greenville, S. C.
WEEKS, H. P.	17.5	Orangeburg H. S.	Orangeburg, S. C.
WESTON, J. B.	12	Charleston H. S.	Charleston, S. C.
WETHINGTON, W.	14	Thomasville H. S.	Thomasville, Ga.
WHEDBEE, S. M.	14	Hertford H. S.	Hertford, N. C.
WILES, G. R.	13.5	Ashboro H. S.	Ashboro, N. C.
WILLIAMS, W. E.	13.1	Fairfax H. S.	Brunson, S. C.
WILLIAMSON, H. C.	16.5	Hastoc Sch.	Florence, S. C.
WILSON, D. I.	14	Rome H. S.	Oaks, S. C.
WILSON, G. P.	12	Sumter H. S.	Florence, S. C.
WOOD, F. C.	13.5	W. F. S.	Kelton, S. C.
WOOTEN, S.	12	S. Ga., exam.	Eastman, Ga.
YOUNG, E. B.	14	Albany H. S.	Albany, Ga.
ZEIGLER, J. G.	17	Bamberg H. S.	Bamberg, S. C.

BATTALION ORGANIZATION.

Lieutenant and Adjutant _____WITSELL, J.
Lieutenant and Quartermaster_____NICHOLSON
Sergeant-Major _____JAMES, W.
Quartermaster-Sergeant _____DOYLE

COMPANY "A"	COMPANY "B"	COMPANY "C"	COMPANY "D"

Captains:
Jones, H.   Alexander, C.   Brown, A.   Sanders, J.

First Lieutenants:
Thompson, F. Coleman, H.   Lawson   Cannon
Wolfe   Cothran

Second Lieutenants:
McGowan   Street   Hagan   Pearlstine
Hane   Reynolds   Wilson, J.   Ford, F.

First Sergeants:
Smith, E.   Middleton   Crawford   Rainsford

Sergeants:
Latimer   Williams, P. Evans   Seymour
Sanders, M.   Padgett   Riley, A.   Hasell
Carrington   Whaley, E.   Allan, G.   Arthur, L.
              Thomas

Corporals:
Lewis   Witsell, A.   Mood   Williamson
Pollock, E.   Wulbern   Poulnot, C. Heinsohn
Platt   James, H.   Hampton   Arthur, H.
Sanders, P.   Cooper   Dotterer   Simons, B.
Morgan, E.   Ross   Dial   Hill
Brice   Whitaker   Kaminer   Jarvis
Tiedemann   Lucas   Langley
Antley

Color Guard:
Sergeants Carrington and Allan, G.; Privates Maxwell and Moore, C.

"Star of the West" Medal, 1918, Cadet Private T. W. Williamson
"Willson Ring", 1918_____Cadet Captain B. R. Stroup
Scholarship Medal, 1918_____Cadet Captain G. G. Cromer
Marksman's Medal, 1918_____Cadet Private G. H. Allan
Prize Company, 1918. "W. C. White" Medal,_____
_____Company "C", Cadet Captain W. R. Mood.

THE CITADEL BAND

Director	CARL H. METZ
Drum Major	WOLFE, LIEUTENANT
Piccolo	FAIREY
Solo B♭ Clarinet	ROGERS, J.
First B♭ Clarinet	AUSTIN
Second B♭ Clarinet	DILTZ
E♭ Saxaphone	HUTCHINS
Solo B♭ Cornet	METZ
Solo B♭ Cornet	FISHBURNE
First B♭ Cornet	CLYMER
First E♭ Alto	REEVES
Second E♭ Alto	WILLIAMS, E.
First Trombone	BECK
Second Trombone	DILL
Valve Trombone	BLOOD
Baritone	WULBERN, CORPORAL
E♭ Tuba	EVARTS
BB♭ Bass	TAYLOR
Snare Drum	FROST
Bass Drum	WANNAMAKER
Cymbals	ZEIGLER

PART II

HISTORICAL SKETCH

Previous to the year 1841, the State of South Carolina had two depositories for its arms and munitions of war—one known as the Citadel, in the City of Charleston; the other known as the Arsenal, in the City of Columbia. These were guarded by companies of enlisted men, and trained officers, and were maintained at an expense of twenty-four thousand dollars per annum.

It was Governor Richardson who made the suggestio that these garrisons be replaced by young men, who, while serving as guard, should receive military training, and instruction in the practical and mechanic arts. Under the administration of his successor, Governor Hammond, an Act of the Legislature was passed, on December 20, 1842, creating the Citadel and Arsenal Academies.

The Board appointed by the Governor to carry out the purpose of the Act lost no time in performing their duty, and the faculty of the Citadel was elected on February 23 following. By the twentieth of March, both the Citadel and the Arsenal were in operation.

In arranging the course of studies for the Citadel, the report of the Board to the General Assembly says:

"The Board have aimed at a system of education at once scientific and practical, and which, if their original design is carried out, will eminently qualify the Cadets there taught for almost any station and condition of life.

"During the course, besides the usual branches taught at the primary schools in the State, they will be instructed in the history of South Carolina, modern history, the French language, every department of mathematics, bookkeeping, rhetoric, moral philosophy, architectural and topographical drawing, natural philosophy, chemistry, geology, mineralogy, botany, civil and military engineering, the constitutional law of the United States, and the Law of Nations. In addition to that course, they will be instructed in the duties of the soldier, the School of the Company and of the Battalion, the Science of War, the Evolution of the Line, and the duties of commissioned officers."

The Arsenal, at first co-equal with the Citadel, was soon incorporated with it, and had for its special function the instruction and training of the recruits forming what was known as the Fourth Class.

The first class, numbering six men, was graduated in 1846. C. C. Tew, the first honor man of his class, and proto-graduate of the institution, afterwards founded the Hillsboro Military Academy, North Carolina, was Colonel of North Carolina troops in the Confederate Army, and was killed at Sharpsburg, September, 1862, while commanding Anderson's brigade.

The value to the State of the military training given at the Citadel is strikingly shown by the fact that, of the two hundred and forty graduates before the close of the War Between the States, about two hundred were officers in the Confederate service, and forty-three laid down their lives upon the battlefield. The list of Citadel officers in that great conflict is an honor roll of which any institution may well be proud.

There are two dates in the history of the State Military Academies which mark the boundaries of this greatest military struggle of the century. Between January 9, 1861, and May 9, 1865, what a tragic history was enacted!

On the first date, Maj. P. F. Stevens, Superintendent, and a graduate of the Citadel, in command of a detachment of Citadel Cadets, manning a battery of 24-pounders on Morris Island, drove off the steamer "Star of the West," which was attempting to relieve Fort Sumter—thus firing the first hostile shot of the War.

On the latter date, Capt. J. P. Thomas, Superintendent of the Arsenal, and also a graduate of the Citadel, with the Cadets of his command, had a skirmish with Stoneman's raiders, near Williamston, S. C., thus firing the last shot of the War delivered by any organized body of troops east of the Mississippi River.

At the present time, the Corps of Cadets has an annual drill for the "Star of the West" medal, a handsome trophy for the best-drilled Cadet of the Corps, presented to the institution many years ago by Dr. B. H. Teague, a veteran of the War. This medal gets its name from a piece of oak wood, in the form of a star, taken from the historic vessel.

At the fall of Charleston, in February, 1865, the Citadel was occupied by Federal troops. The Corps of Cadets was at that time in the field in the upper part of the State, and never returned to the institution, which continued in the hands of the United States military authorities, in spite of the best efforts of the State to recover it, until 1881.

In that year, Governor Hagood said in his annual message to the General Assembly:

"The State Military Academy at Charleston has been suspended in its operations since the late Civil War. This has been due to the fact that the building known as the Citadel, in which it has its seat, has been since the close of the war in the possession of the United States authorities, and has been used by them until recently as a military post.

"It is understood that the General Government is now prepared to restore it to the custody of the State. It is desirable that this property be recovered, and again devoted to the purpose of higher education, in the facilities for which our needs are greater than the source of supply.

"A measure will probably be submitted to you at this session to accomplish this purpose and you will permit me to say that, in my judgment, it is now practicable to reopen the school, and it ought to be done."

An Act to authorize the reopening of the South Carolina Military Academy was passed by the General Assembly, and approved January 31, 1882.

The Citadel was reopened on October 1, 1882, with 189 Cadets, and has been in continuous and successful operation ever since.

In 1888, the sum of $77,250 was recovered from Congress for the occupation of the Citadel building by the Federal troops, and for the destruction by fire of the West Wing while occupied by them. With this fund the Wing was restored, and the building thoroughly equipped in its departments.

In 1908, the Central Police Station, which was erected on the King Street end of the Citadel property just after the earthquake in 1886, was purchased by the State, and fitted up for the use of the college. In February, 1910, an appropriation was made by the Legislature for the addition of a fourth story to the main building, for cadet barracks. This addition was constructed during the summer of 1910, and the capacity of the institution increased to three hundred cadets.

In 1911, the Legislature made an appropriation of fifty thousand dollars for the construction of the

Meeting Street Extension, which completed the design, and furnishes equipment for all anticipated needs of the institution for the future.

In this year, also, the Legislature fixed the title of the institution as "The Citadel, The Military College of South Carolina."

LOCATION

The Citadel is situated in the center of the City of Charleston, a city noted for its culture and refinement, and full of associations dating from the earliest times of American history.

The climate of Charleston is mild and healthful, being free from the rigors of the severe winters experienced further north and in the interior cities, and tempered in summer by constant sea-breezes.

The Citadel fronts on Marion Square, the largest and one of the most beautiful open squares in the city, and the drill and parade grounds of the Corps of Cadets.

BUILDINGS

The Citadel buildings consist of a main central building, the East and West Wings, the King Street Extension, the Meeting Street Extension, and the Gadsden Gymnasium. The Cadets are quartered in the Main Building, which is built in the form of a rectangle surounding a large interior court called the Quadrangle, where the ordinary formations of companies and classes are made. The second, third, and fourth floors of this building are devoted almost exclusively to the dormitories of Cadets, and are known as Cadet Barracks.

The King Street Extension is a large three-story structure, containing on the first floor the Mess Hall and Kitchen, on the second the Drafting Hall, and on the third floor, the Infirmary.

The Meeting Street Extension contains the artillery and infantry armories, a large athletic hall, band-room, halls for the Calliopean and Polytechnic Literary Societies and the Cadet Y. M. C. A., class-rooms, officers' quarters, and the astronomical observatory.

The East Wing contains the Chapel, the chemical, physical, and geological laboratories, and the laundry.

The West Wing is reserved entirely for officers' quarters.

The guard-room, reception-room, and library are on the first floor of the Main Building. The Gymnasium Building lies between the Main Building and the West Wing, and is reached from the Quadrangle through the west sallyport.

The entire group of buildings is heated by a hot water system, which guarantees a uniform winter temperature of seventy degrees throughout; and is lighted by electricity.

ACADEMIC YEAR

The Academic year begins September 20, and ends the fifteenth of June.

It is divided into three terms of three months each, and examinations are held at the end of each term, after which reports showing the record of the cadets in studies and conduct are sent to parents and guardians.

MILITARY SESSION

A military session, devoted exclusively to military instruction in camp and field work, is held usually at the end of the third term. Sometimes a practice march of two weeks is held at the close of the second term instead.

The annual target practice is held in the Spring, and a marksman's medal, to be worn for a year, is awarded to the cadet making the best record.

Following is a list of the various encampments which have been held:

1889—Greenville, S. C.
1891—Spartanburg, S. C.
1892—Fort Moultrie, Sullivan's Island, S. C.
1893—Aiken, S. C.
1894—York, S. C., march to King's Mountain battlefield, and return.
1895—Camden, S. C., march from Columbia to Camden.
1896—Sumter, S. C., visit to battlefield of Eutaw Springs.
1897—Anderson, S. C. march to Clemson College, and return.
1899—Orangeburg, S. C.
1901—Darlington, S. C.
1903—Rock Hill, S. C., march to Indian Shoals Power Dam and return.
1904—St. Louis Exposition.
1905—Columbia, S. C.
1906—Practice march to Pinopolis, S. C., and return.
1907—Jamestown Exposition.
1908—Practice march to Walterboro, S. C., and return.
1909—Coast Artillery fortifications, Sullivan's Island, S. C.
1910—Greenwood, S. C.

1911—National Guard Range, near Charleston.

1912—National Guard Range, near Charleston.

1913—Coast Artillery fortifications, Sullivan's Island, S. C.

1914—Practice march to Orangeburg, S. C.

1915—Mount Pleasant, S. C.

1916—Mount Pleasant, S. C.

1918—Plattsburg, N. Y.

FURLOUGHS

A suspension of Academic work for ten days, including Christmas Day and New Year's Day, is required by law.

The Corps of Cadets is furloughed from Commencement Day, in June, until the opening of the following session on September 20. There are no Easter holidays.

During the session, furloughs will be granted to Cadets only for some urgent reason, and parents are requested not to apply for leaves of absence for their sons unless in case of necessity.

Parents applying for the admission of their sons to The Citadel relinquish control over them to the authorities of the institution. The time which is allotted to studies and military work is essential for the completion of the required courses, so that cadets must not be absent except when it is absolutely necessary.

SYSTEM OF MANAGEMENT

The Citadel is essentially a military college, and all the students live in Cadet Barracks, under the same discipline as in use at West Point. From reveille to taps, the Citadel Cadet passes a full and busy day, every hour of his time being accounted for, and its appropriate task performed.

The regular habits of study and living thus formed, the attention to duty, obedience to authority, and love of order inculcated, are considered among the most valuable features of the education given. While few of the graduates of the college enter the military profession, hundreds in all the walks of civil life attest to the high value of the training they received at the institution.

RELIGIOUS SERVICES

The religious training of Cadets is provided for by daily chapel services attended by the Corps, and by attendance on Sunday mornings at the services in the various churches of the city. The institution being non-sectarian, the Cadet companies are assigned in rotation to the various Protestant churches, but occasionally special individual leave is granted Cadets who are communicants, to attend the services and commune at some church of their particular denomination. Cadets of the Roman and Hebrew faiths are excepted from the above rule when the request is made, and form special squads which attend only the services of their own faith.

The military system works not only for the good health but for the morals of the Cadets. Moreover, the barracks life of the students precludes much waste of time or loss of character.

Cleanliness, temperance, regularity, and courtesy are insisted upon, and personal responsibility and a high sense of honor are stimulated and developed.

LIBRARY AND READING ROOM

The Library contains a large number of books of reference for all the departments of the College, besides being well supplied with works of standard and current fiction. It is catalogued, and in charge of a competent librarian.

The reading room is a large, airy and well-lighted room on the the ground floor, directly accessible from the Quadrangle, and is supplied with the following magazines:

Aerial Age	McClure's
Arms and the Man	Life
Army and Navy Journal	Literary Digest
Atlantic Monthly	Munsey's
Century	Nation
Colliers	New York Times
Confederate Veteran	North American Review
Cosmopolitan	Outlook
Everybody's	Physical Culture
Flying	Popular Mechanics
Forum	Popular Science Monthly
Geographic Magazine	Review of Reviews
Harper's	Saturday Evening Post
Hearst's	Scientific American
Illustrated London News	Scribner's
Independent	The State
Leslie's	World's Work

LITERARY SOCIETIES

There are two Literary Societies in the Corps—the Calliopean, organized in 1845; and the Polytechnic, two years later. These Societies are officered and controlled by the cadets. They have comfortably furnished halls, and hold meetings on Saturday night of each week. Frequently during the session, upon the invitation of the Societies, men of note deliver addresses before the Corps and their friends upon literary topics.

The Societies of the Citadel send a representative each year to the oratorical contest of the South Carolina Oratorical Association, which is composed of all the colleges of the State; they also engage in an annual joint debate with the societies of the College of Charleston.

While these societies are strictly in the hands of the Cadets, the work done by them is considered second to that of no department in the institution in its educational value.

Y. M. C. A.

The Citadel Y. M. C. A. was organized in 1886, and is in flourishing condition. In the past five years, it has grown to be one of the strongest student organizations in the State.

While the faculty has oversight of the work, it is an institution "of the boys, for the boys, and by the boys." Here all meet on equal terms; the old men are welcomed back, and there is thrown around the new student a wholesome and affectionate atmosphere, which enables him to meet more courageously and successfully the hard battles that every student must fight.

The supervision and extension of the work is in the hands of a Student Committee, Cabinet, and Executive Secretary. The latter is employed by the Student Department of the local Association, and devotes his entire time to work among the schools and colleges of the City.

The regular weekly meetings of the Association are addressed by prominent clergymen, professional and business men of the city. At frequent intervals, these meetings are addressed by noted men from the greatest religious, secular, and educational institutions of the country, speakers brought here by the city organization. Contact with these men is an education in itself, and the moral and religious influence of their helpful messages is most profound. A series of "Vocational Talks" finds place among the lectures delivered at these meetings.

The efficiency of Association leaders, and the effectiveness of the work, is greatly increased by the yearly attendance of Cadets at County, State, and International Conventions and Summer Schools. Bible Study Classes are organized each year, and taught by members of the Faculty and the Senior Class, and the religious leaders of the city. In 1913-1914, ninety per cent. of the student-body enrolled in these classes. Mission study is also provided for, and from time to time the students' horizon is broadened by lectures of Foreign Work representatives. The religious and economic problems of the homeland are also given due consideration.

Two of the College publications are under direct control of the Association. A *News Notes* is published monthly, containing all important College news. A *Handbook*, published by the Association, is presented each year to the students at the beginning of the first term.

The social side of the students' life is provided for by frequent social affairs, and by the Social rooms. These rooms are provided with pool tables, an inner-player piano, game tables, a library, etc. The rooms are self-sustaining, and meet a real need of the students.

The Faculty heartily commends the work of the Association, and it is recommended that every parent or guardian encourage the student under his care to affiliate himself with the Y. M. C. A. as soon as he enters The Citadel.

ATHLETICS.

The climate of Charleston permits open-air exercises throughout the year; and the setting-up exercises, Butt's Manual, daily drill, etc., on Marion Square, and the sports of the Cadets, furnish the best means of securing bodily health and growth; but special attention may be given to those Cadets who may need particular exercises for specific needs.

The coaches of the football and baseball teams are competent and experienced and use due care to prevent injurious training on the part of Cadets, who engage in those sports.

It is the policy of the institution to give every reasonable encouragement to athletics, and to see that all contests are conducted on a clean, amateur basis. The athletic interests are controlled by a well-organized cadet athletics association, under the supervision of a Faculty athletic committee. The Citadel is a member of the Southern Intercollegiate Athletic Association, and it is represented at the annual track and field meet of this organization.

A wide range of opportunity is offered to the Cadet to find some branch of athletics in which he may excel or from which be may derive pleasure and profit. Football, baseball, track and field sports, tennis, basket-ball, and relay racing are the branches in which regular teams are organized. The danger of athletic sports is minimized by having the men under the direct care of competent coaches, and by strict examinations by the surgeon.

While athletics are encouraged and supported by the authorities, they are firmly subordinated to the prescribed work of the College, and no interference that will materially hamper the progress of the Cadets in their studies, or introduce irregularities into the rigid routine of the College, is permitted.

It is the observation of the authorities that clean, well-conducted athletics in an institution of learning foster a fine spirit of loyalty and manliness. It is the purpose here to derive a full measure of this benefit, and in addition wholesome recreation to the Cadets.

ASSOCIATION OF GRADUATES.

The Association of Graduates was organized in 1877, since which time it has been an active agent in promoting the best interests of the alma mater. The annual meeting and supper are held at the Citadel, at the time of the Commencement Exercises, in June.

The *Bulletin*, a quarterly publication issued by the Association, keeps the graduates and ex-cadets informed of the principal transactions at the College, besides containing many notes of interest about the alumni.

OFFICERS OF THE ASSOCIATION OF GRADUATES OF THE CITADEL, 1918.

D. G. DWIGHT, *President*, Charleston, S. C.
F. B. GRIER, *First Vice-President*, Greenwood, S. C.
S. P. ANDERSON, *Second Vice-President*, Charleston, S. C.
W. W. SMOAK, *Third Vice-President*, Walterboro, S. C.
S. L. REID, *Secretary*, Charleston, S. C.
T. P. LESESNE, *Treasurer*, Charleston, S. C

DIRECTORS.

J. G. PADGETT, Walterboro, S. C.
W. S. LEE, Charlotte, N. C.
D. C. PATE, Bennettsville, S. C.
F. M. ELLERBE, Jonesville, S. C.
A. P. MCGEE, Charleston, S. C.

EXPENSES.

The Citadel is a State institution, and is not maintained for profit. Only the cost of supporting the student is required.

The fees, exclusive of uniforms, are $250 a year, and will include the cost of board, tuition, laundry, lights, heat and hospital. This amount will be paid in three instalments, the first on September 20th, $90, and the other two payments on January 1st and April 1st, $80 each.

A cadet discharged during any term will not be entitled to a refund of any of the amount paid for maintenance for that term.

The rates above do not include the cost of uniforms or books. A deposit of $10 to cover cost of books, $5 for breakage fee, and $95 for uniforms must be made at the beginning of the first term on September 20th. Owing to the uncertainty in the prices of uniforms the above amount is only approximate, and includes an olive drab uniform estimated at $30, a full dress uniform $40, an overcoat $25. If the prices should be reduced, full credit will be given the cadet for any amount which may be saved on uniforms. It is also expected that the U. S. Government will furnish a considerable portion of the uniforms worn by cadets, in which case due credit will be allowed, and the unexpended balance of a cadet's clothing account will be returned to him at the end of the session.

The first payment required to be made on September 20th will, therefore, be $200. The other two payments, due January 1st and April 1st, will be $80 each.

A remission of forty dollars for tuition is allowed to residents of the State of South Carolina, when acceptable certificates showing inability to pay are filed with the State Board of Charities, as required by law.

Beneficiary cadets are required to make a deposit the first year of $40 to cover the cost of overcoat, books, and breakage fee.

All cadets are required to furnish their own bed-clothes. Bedsteads and mattresses are provided at The Citadel, but pillows are not furnished. Cadets are also expected to come provided with underclothes. A list of these articles will be found elsewhere in the catalog.

There are no extras charged for at The Citadel. The hospital facilities are excellent, and all ordinary cases of sickness are treated by the physician and nurses of the college without expense to the parent. Surgical cases, however, requiring the removal of the student to the Infirmary in the city, must be paid by the parent or guardian; and also special treatment of eyes or ears, dental services, etc.

To those pay cadets whose tuition is remitted, the last two payments are $60 each.

Besides the fees explained above, parents are expected to make their sons a reasonable, but not extravagant, allowance of pocket money.

All cadets are expected to take an interest in the athletic games, and in certain social functions of the Corps of Cadets. They will wish to attend all the football and baseball games played by the cadet teams in Charleston; and also bear their share of the cost of the social functions. Besides the mess-hall dances, musicales, and lectures, five formal dances are annually given: the Thanksgiving Hop, the Christmas Hop, the Senior Hop, the Annual Picnic, and the Commencement Hop.

A system of Class funds, under the regulation of a Cadet Council and the Commandant of Cadets, has been established, whereby a cadet can participate in all the college activities at a minimum of cost.

It is very strongly recommended to parents that they send with the first instalment of dues a deposit of twelve dollars, which will entitle their son to attend all athletic games and social functions, and also give him membership in the local Y. M. C. A., where he will have the many advantages usually offered by this excellent organization.

Cadets will not need any large amounts of pocket money. They have leave in the city only on Friday nights and Saturday and Sunday afternoons, and there should be no occasion for them to spend any large amount of money. They do need, however, a small sum (from one to two dollars per week). The Quartermaster cannot take care of these allowances but they should be sent directly to cadets, either by check or postoffice order. They should never be sent in cash, nor in large amounts. A small, regular weekly allowance, on which the cadet can count, and by which he can regulate his expenses is the best way to cultivate in a student proper habits of economy.

To recapitulate:

The cost of supporting a cadet one year at The Citadel may be estimated as follows:

Fees, covering board, lodging, tuition, laundry, lights,
  heat, hospital, books, breakage_____$265.00
Uniforms, estimated _____ 95 00
Class Fund _____ 10 00
Pocket money, about _____ 40 00

    Total_____$410 00

To this amount should be added the railroad fare to and from Charleston, and a small amount the first year for shoes and underclothes. After the first year, a cadet's uniform account can cover underclothes and shoes if he is at all careful, and the amount above can be reduced by at least thirty dollars.

All remittances should be made to MAJ. E. M. TILLER, QUARTERMASTER, THE CITADEL, CHARLESTON, S. C.

LIST OF ARTICLES WHICH EACH CADET MUST BRING TO THE CITADEL WITH HIM.

Six white or negligee shirts.
Six summer undershirts.
Four winter undershirts.
Three nightshirts, or pajamas.
Twelve linen collars, straight, white, one and three-quarters inches high.
One black tie.
Six pairs cuffs, white linen.
Six summer drawers.
Four winter drawers, or union suits.
Six pairs of black socks.
Six handkerchiefs.
Six towels.
One clothes bag.
One pair high, laced, black leather shoes (patent leather is not permitted).
One pair high, laced, tan leather shoes, army pattern.
One clothesbrush, hairbrush, toothbrush, and comb.
One pillow.
Three pillowcases.
Four sheets for single bed.
Two blankets.
One comfort or spread.
One bathrobe.
One sweater.

Clothing to be marked as follows: Sheets, towels, and handkerchiefs in the corner; pillowcases in corner, at open end; collar and cuffs on inside, near center; shirts on band, near buttonhole at back of neck; undershirts and nightshirts on piece of cloth containing buttonholes, inside, near upper front; socks on leg, near top.

PART III

REQUIREMENTS FOR ADMISSION

Applicants for admission must be not less than sixteen nor more than twenty years of age. They must be at least five feet high, and physically able to do military duty.

Applications must be made by parents or guardians to the Superintendent, and should be accompanied by a certificate from the principal or president of the school or college which the student last attended.

1. The requirements for admission to the Fourth, or Freshman, Class at The Citadel are:

Fourteen High School units, of which two and a half should be in Mathematics, three in English, two in History, and one in Science. The others may be selected from any given in the list of Standard High School units.

Diplomas from High Schools whose courses cover these requirements will admit the applicant without examination. The certificate from a High School which offers only twelve units will be accepted for conditional entrance. But in no case will a certificate covering less than twelve units be accepted for admission. Applicants not having a diploma or a satisfactory certificate will be examined in such studies as will show their proficiency in the studies covering twelve units. Applicants are requested not to send their diplomas, but to have the certificate from the back of the catalog filled out and send in with application.

It is the policy of The Citadel to discourage students from coming from any community that maintains a four-year High School course until they have completed the fourth year.

2. The competitive examinations for Beneficiary Scholarships will be based as nearly as possible on the first three years' work of the High Schools.

3. The following information concerning the scope of these examinations is furnished prospective candidates.

MATHEMATICS.

The examination in Algebra will cover the operations through quadratic equations to be found in any text-book in common use, and will lay particular stress upon factoring, solution of simple simultaneous equations, square and cube root, theory of exponents, and radicals. In Plane Geometry, the examination will be given on the elementary propositions.

ENGLISH.

ENGLISH GRAMMAR—The examination in this subject will include spelling, punctuation, the various constructions of the parts of speech, and the analysis of the English sentence.

ELEMENTARY RHETORIC—The examination in this subject will be based upon the use of words in the sentence, the structure of the sentence, and the various methods of developing the paragraph. In addition to this, the applicant must have a fair knowledge of narration, description, and letter-writing.

LITERATURE—It is expected that the applicant will be familiar with some of the best literature, both in prose and poetry, and with the lives of the authors. The following is suggested as a suitable list of works to be studied: Shakespeare's "Macbeth"; Macaulay's "Life of Johnson," or "Hasting's;" Gray's "Elegy in a Country Churchyard"; Goldsmith's "The Deserted Villiage"; Irving's "The Sketch Book"; Franklin's Autobiography; George Eliot's "Silas Marner"; Simms, "The Yemassee"; Timrod's War Lyrics.

HISTORY.

HISTORY OF THE UNITED STATES—The candidate should be prepared to name the European countries that took part in exploring and settling North America, and to give an account of the founding of the principal colonies in what is now the United States. He should be able to tell what the chief occupations of the people in those colonies were, to give an account of the colonial wars, and to discuss the troubles of the English colonists with the mother country. He should know how the causes and results of the principal wars to which the United States has been a party, and be able to give a chronological account of the chief events of each. He should be prepared to show a knowledge of social and industrial changes, more especially those of the last fifty years, and to name the Presidents in order, and give a discussion of one or more important events in the administration of each. Some such text as Hart's *Essentials in American History* is recommended.

ANCIENT HISTORY—The candidate should be able to give a chronological account of the rise and fall of the various oriental nations, and to explain what they contributed to the European world. He should familiarize himself with social and political conditions and changes in ancient Athens and Sparta; and should be able to show knowledge of the culture of the ancient Greeks; and to indicate what the modern world owes them. The period from the opening of the Persian wars through the break-up of Alexander's Empire should be thoroughly studied. In Roman History, the candidate should be able to give an account of the governmental changes in the Roman world, to the founding of the Empire, and should be able to tell how, Rome expanded over Italy and then over the Mediterranean world. He should be able to name the principal emperors, and give an important event in the reign of each, and to explain the decay and fall of the Empire. The text by Myers is recommended.

PHYSICAL GEOGRAPHY.

The examination on this subject, besides showing the candidate's knowledge of descriptive Geography, is designed to elicit his knowledge concerning the main facts of air and earth sciences. An intelligent perusal of any standard text, such as Tarr's, Radway's, or Maury's, with special attention to the scientific terms employed, will furnish all necessary information.

The following may be considered a general outline of the subject.

- I. The Universe—Nebular Hypothesis, Solar System, The Sun, Planets, Satellites.
- II. The Earth—surface, movement.
- III. The Atmosphere—General Features, Light, Sun's Heat, Variations of Temperature, Winds, Storms, Moisture, Climate.
- IV. The Ocean—General Characteristics, Movements.
- V. The Land—Earth's Crust, Wearing, River Valleys, Glaciers and Glacial Period, Seas and Lakes, Plains, Plateaus, Mountains, Volcanoes, Earthquakes, Geysers.
- VI. Relation between Range of Plant and Animal Life—Geographical Distribution of Labor Dependent on Physical Geography.

STANDARD HIGH SCHOOL UNITS.

ENGLISH:

1—Higher English Grammar and Grammatical Analysis____1
2—English Composition and Rhetoric _____1
3—Critical Study of Specimens of English Literature_____2

MATHEMATICS:

1—Algebra to Quadratic Equations_____1
2—Algebra—Quadratics, Progressions, and Binomial Theorem _____ ½
3—Advanced Algebra, including Permutations and Combinations, Determinants, and Numerical Equations____ ½
4—Plane Geometry _____ 1
5—Solid Geometry _____ ½
6—Plane Trigonometry _____ ½

LATIN:

1—Grammar and Composition, of First Book_____1
2—Caesar, Books I-IV_____1
3—Six Orations of Cicero_____1
4—Virgil's Æneid, first six books_____1
5—Cornelius Nepos, first fifteen Lives_____1

HISTORY:

1—Greek and Roman History_____1
2—Medieval and Modern History _____1
3—English History _____1
4—American History and Civics _____1

SCIENCE:

1—Physiology, with field and laboratory work_____1
2—Experimental Physics _____1
3—Physiology, with laboratory work_____ ½
4—Inorganic Chemistry, with laboratory work_____1
5—Botany, with laboratory work_____1
6—Zoology _____1

GREEK:

1—Grammar and Composition _____1
2—Xenophon's Anabasis, Books I-IV_____1

GERMAN:

1—Half of Elementary Grammar, and 75 pages Reading____1
2—Elementary Grammar completed, and 150 pages Reading__1

FRENCH:

1—Half of Elementary Grammar, and 100 pages Reading__1
2—Elementary Grammar completed, and 200 pages Reading 1

SPANISH:

1—Half of Elementary Grammar, and 100 pages Reading__ 1
2—Elementary Grammar completed, and 200 pages Reading 1

COMMERCIAL SUBJECTS:

1. Commercial Arithmetic _____1
2. Bookkeeping _____1
3. Shorthand _____1½
4. Typewriting _____½
5. Commercial Geography _____1
6. Agriculture _____1 to 4

MANUAL TRAINING.

1. Free-Hand Drawing_____1
2. Mechanical Drawing _____1
3. Shopwork _____1 to 2

DEPARTMENTS OF INSTRUCTION

SYNOPSIS OF COURSES

FRESHMAN YEAR—*Required of all Students.*
Mathematics 1; Physics 1; English 1; History 1; French 1; Military Science 1.

SOPHOMORE YEAR—*Required of all Students.*
Mathematics 2; Chemistry 1; English 2; History 2; French 2; Drawing 1; Military Science 2.

JUNIOR YEAR.

Required: English 3; French 3, German 1, or Spanish 1; Military Science 3.
Electives: (1) Engineering Course—Mathematics 3; Engineering 1; Physics 2; Drawing 2.
(2) Chemistry Course—Chemistry 2; Chemistry 3; one other junior course.

SENIOR YEAR.

Required: German 2, or Spanish 1; History 4; Military Science 4.
*Electives*s (1) Engineering Course—Mathematics 3; Engineering 3; Drawing 3.
(2) Electrical Course—Physics 3; Physics 4.
(3) Chemistry Course—Chemistry 4; Chemistry 5.
(4) Chemistry-Physics Course—Chemistry 4; Physics 3.
(5) Biology Course—Chemistry 4; Biology 1.
(6) English Course—English 4; one other senior course.

DEPARTMENT OF MATHEMATICS AND ENGINEERING.

Major Thomas
Major Le Tellier
Captain Hair
Lieutenant Myers

The method of instruction in this department is by text-book and recitation, supplemented by lectures and field work. Much of the history and philosophy of the various branches is incidentally given by lecture. It is sought to make the subject interesting as well as instructive. The aim is, primarily, to draw out and develop the powers of the student, to train his faculties rather than to cram his mind with information undigested and not assimilated.

The course in Mathematics for the first two years is required of all students. Engineering with Calculus is elective the last two years.

MATHEMATICS

Course I. First Year. Required_____5 hours per week

Capt. Hair—Lt. Myers.

Algebra, Solid Geometry, Plane Trigonometry.

The requirements for entrance being Algebra through quadratic equations and Plane Geometry, after some review of the elements, Advanced Algebra is completed. In Trigonometry, special attention is given to the solution of triangles, and to other applications in courses to follow. Practice in the use of logarithms is required. The text used is Wentworth's Plane and Spherical Trigonometry.

For Solid Geometry, which is completed, the text-book is Ford and Ammermann's Solid Geometry.

Course II. Second Year. Required_____3 hours per week

Major Thomas—Capt. Hair.

Spherical Geometry and Trigonometry, Analytic Geometry.

The parts of Spherical Trigonometry essential to the courses that follow are given in 4 weeks; then the remaining time is given to the Analytic Geometry. In this, while the straight line and conic sections receive as usual the most study, some of the higher curves are studied, and due attention is given to planes and solids. The analytic method is emphasized, and employed in the solution of practical problems. The text-book for the class is Wilson and Tracey's Analytic Geometry.

Course III. Third year. Elective_____3 hours per week.

MAJOR THOMAS.

Differential and Integral Calculus.

The methods commonly used in the Calculus are presented, and the advantages of each method made apparent, while the method of limits is taken as a foundation. The subject is taught not as pure theory alone, but the student is made to realize what an efficient means it is of treating practical cases in Engineering and Physics.

N. B.—The text studied by the class is Applied Calculus.

CIVIL ENGINEERING

Course I. Third Year. Elective_____6 hours per week

MAJ. THOMAS—MAJ. LE TELLIER.

Plane Surveying, Drawing, Roads and Pavements.

The instruction in Surveying is practical and theoretical. Surveys are made with Compass, Transit, and Level; and areas are computed by each student. Stadia work is done, and the uses of the Plane Table and the Sextant are shown. Determinations of the True Meridian are made by observations on Polaris and on the sun. The theory and practice of laying out railway curves and of calculating earth work is included. Simple triangulation and topographical and hydrographical surveying are treated. The drawing is described in the Drawing Department. A short course is given on the construction and maintenance of Roads and Pavements. Recitations 2 hours a week, field work 2 hours, and drawing 2 hours a week.

Course II. Fourth Year. Elective_____3 hours per week
MAJOR THOMAS.

Analytic Mechanics, Mechanics of Materials.

This course in Theoretical and Applied Mechanics is on the Principles of Mechanics and their application to structures.

The theory of central forces and its application to the motion of the planets are given. The stresses in beams, columns, and shafts, and in the simple forms of bridge and roof trusses are studied.

The importance of Mechanics as the basic study for Engineering is recognized.

Course III. Fourth Year. Elective_____3 hours per week.
MAJOR THOMAS.

Hydraulics, Sanitary Engineering.

In Hydraulics, the study is on the pressure of water against dams and other structures, and on its flow in pipes, rivers, and canals. The study of air pressure and that of steam is included with other fluids.

The Sanitary Engineering includes water supply, sewerage, garbage disposal, and other means of preserving the health of the community. Attention is given to vital statistics, and the importance of their systematic collection. The germ theory of disease is treated, and the specific bacteria described.

There is a supplementary course in Drawing, described under Drawing Department, that is required with Course II, III, in Engineering; and while Course III, in Mathematics, is elective to all students, it is required before these two courses in Engineering can be taken. The latest and best text-books suitable are used by the students in Engineering, and various standard works on the subject are in the Library, available for reference. The attention of the students is called to the latest practice as set forth in *Engineering News.*

While the instruction is by text-book and recitation, with practice field work, lectures explanatory and supplementary to the text are daily given.

The equipment for instrumental work in the field includes Transits and Levels of the latest improved kind, Compasses, Plane Table, and Sextant.

DEPARTMENT OF ENGLISH.

MAJOR MCGILLIVRAY
LIEUTENANT VINING.

The object of this course is to train students in the correct use of their mother tongue, and to give them a fair knowledge of its literature, both English and American.

The work of the course is required in the first three years, and is as follows:

FRESHMAN CLASS. Three hours a week—Lieutenant Vining.

A. TYPES OF LITERATURE—One hour a week. The various literary types are closely studied in representative selections, viz.: a group of short stories, a novel, a (Shakespeare) play, an essay, and selected poems.

B. RHETORIC—Two hours a week—Lieutenant Vining.

This subject includes a thorough review of the parts of speech, their inflections and their uses, with a close study of syntax and logical analysis. The aim of the course is to give the student a practical command of the English sentence, and much time, therefore, is given to writing compositions. The practice work is confined to narration and description. In addition to the daily written exercises, weekly themes on familiar topics are assigned. Special stress is laid upon spelling, punctuation, capitalization and errors in the use of English.

Text-books. Canby and Opdycke's "Composition"; "The Century Handbook."

SOPHOMORE CLASS. Three hours a week—Major McGillivray.

A. LITERATURE.—The history of English Literature is studied, from the Age of Chaucer to the close of the Victorian Age. Along with the historical setting of each period, and the biographies of the various writers, the course includes the study of at least one typical work of each principal author. In the study of the more important writers, a fair amount of parallel reading is required.

Text-book: Snyder and Martin's "A Book of English Literature."

B. RHETORIC.—One hour a week. The work in this class is a continuation of that begun in the Freshman Class. Exposition and argumentation are studied, and their principles applied in fortnightly themes.

JUNIOR CLASS—Three hours a week.—Major McGillivray.

SHAKESPEARE—This course embraces (a) lectures on the Elizabethan Stage, (b) lectures on Elizabethan English, (c) a careful study of the principal plays, (d) parallel assignments from the less significant.

SENIOR CLASS (Elective)—Three hours a week—Major McGillivray.

THE ENGLISH ESSAY—A careful study of representative essays and essayists from Bacon to Arnold is attempted. De Quincy, Macaulay, and Carlyle are stressed. Parallel assignments.

DEPARTMENT OF HISTORY.

MAJOR MOORE.
CAPTAIN WILLIAMS.

Course I. Required of all Freshmen. History of Modern Europe, 1500-1815. Three hours a week.

Course II. Required of all Sophomores. History of Modern Europe, 1815-1914. Three hours a week.

Course III. Required of all Juniors.

    (a) American Government and Politics. Three hours a week, 4 months.

    (b) Economics. Three hours a week, 5 months.

Course IV. Required of all Seniors. The Governments of Europe and the issues of the War. Three hours a week.

In Courses I, II, and III, as well as in Course IV, the World War will be kept constantly in mind, and a sustained effort will be made to find its roots and to understand its issues. In all Courses, the problems of reconstruction will receive such consideration as under the varying conditions existing in the different classes is logical and feasible.

Course I. This course is based on Volume I of Hayes' "A Political and Social History of Modern Europe." The text is followed closely, its contents being taken up in order. The following, which is a brief outline of the text, is the best general outline that can be given of the course.

1. The Countries of Europe at the beginning of the Sixteenth Century.
2. The Commercial Revolution.
3. European Politics in the Sixteenth Century.
4. The Protestant Revolt and the Catholic Reformation.
5. The Culture of the Sixteenth Century.
6. The Growth of Absolutism in France, and the struggle between the Bourbons and the Hapsburgs, 1589-1743.
7. The triumph of Parliamentary Government in England.
8. The World Conflict of France and Great Britain.
9. The Revolution within the British Empire.
10. The Germanies in the Eighteenth Century.
11. The Rise of Russia and the Decline of Turkey, Sweden, and Poland.
12. European Society in the Eighteenth Century.
14. The French Revolution.
15. The Era of Napoleon.

Throughout the course everything touching English and American History is emphasized. Parallel reading is recommended in connection with the course, but not required. Constant emphasis is placed on chronology; and map-work is regarded as indispensable, eighteen maps being required during the session.

Course II. This course is a continuation of Course I, and is based on Volume II of Hayes' "A Political and Social History of Modern Europe." In this course the text is not followed as closely as in the preceding one; however, the best general outline of the course is that of the text, which is as follows:

1. The era of Metternich, 1815-1830.
2. The Industrial Revolution.
3. Democratic Reform and Revolution.
4. The Growth of Nationalism, 1848-1871.
5. Social Factors in European History, 1871-1914.
6. The United Kingdom of Great Britain and Ireland, 1867-1914.
7. Latin Europe, 1870-1914.
8. Teutonic Europe, 1871-1914.
9. The Russian Empire, 1855-1914.
10. The Dismemberment of the Ottoman Empire, 1683-1914.
11. The new Imperialism and the spread of European Civilization in Asia.

12. The Spread of European Civilization in America and in Africa.
13. The British Empire.
14. International Relations (1871-1914), and the Outbreak of the War of the Nations.

As in Course I, emphasis is placed on everything touching English and American History, and the same plan as to parallel reading is followed. The emphasis on chronology and geography is kept up, twenty maps being required in this course.

Course III. This course falls into two parts: (a) American Government and Politics, and (b) Economics. Endeavor is made to so give the course that the first part constantly looks forward to the second, and the second back at the first; and it is believed that the connection established abundantly justifies the course.

In studying Government and Politics, the students first center their attention on the origin, growth, and operation of our national government and institutions, and later on State and local government and institutions. The historical setting is kept prominently in view throughout the course, it being believed that to understand our institutions one must look to the past, out of which they have emerged. At every important stage, the views of opposing leaders are analyzed and weighed.

In the study of our government, it is sought to understand not only its book theory, but the practical workings as well. The rise, growth, machinery, and place of political parties in the United States are studied. The place of the political party in England is also understood, and a comparison is made between the party government of the United States under an instrument that contemplated no party alignments, and that of England under a scheme of government in which the political party is fundamental.

After a general survey of State government, the several constitutions of South Carolina are briefly considered in their order of adoption. Attention is given to the character of the adopting bodies; and the reasons for the changes made by each are sought for.

The nature of the County and Township are understood; and a close study is made of City government, a critical comparison being made of the aldermanic and commission forms. Attention is given to municipal functions.

The work in Economics is based on Ely's Outlines of Economics, the text being followed closely. Effort is made to make the work practical, but lack of time makes it impossible to go, to any practical extent, into any of the fields in which economic principles apply.

Stress is laid on the importance of wide reading in connection with all work of the Senior Class in this department.

Course IV. The governments of England, France, and Italy will be studied, as will those of Germany and Austria-Hungary. While in a measure this will be a study in comparative government as such, in the main, the study will have for its purpose the understanding of the political principles and ideals of the Allies of America in the World War, and of the conflicting political principles and ideals of Germany and her chief ally. The course will be at one and the same time a course in comparative government and a course in war issues.

DEPARTMENT OF CHEMISTRY.

MAJOR KNOX
LIEUT. CLARKE.

This department embraces three laboratories, two lecture rooms, and store-rooms. The department is amply provided with chemicals and apparatus requisite for the maintenance of a high degree of efficiency in the subject.

The total amount of work offered in the subject consists of five standard college courses, as follows:

Course I. Lectures, two per week; Laboratory, two hours per week. This course is required of all Sophomores.

No previous knowledge of chemistry is assumed, the course being essentially a beginner's course. The student is provided with a text-book and laboratory manual, and these are supplemented with demonstrated lectures and classroom discussions, it being the object of the department to be assured that the student's knowledge is accurate, and his inferences and deductions well founded.

The course gives the cadet that knowledge of chemistry required of all well-educated men.

The economics of the course follows: The commoner elements, their modes of preparation and properties, are first studied; and, as the student grows in breadth of detail, the underlying theoretical conceptions are carefully introduced and emphasized. Great stress is laid on the ionic hypothesis, and its application to the explanation of all such phenomena as double decomposition, hydrolysis, and the completion, in one direction, of reversible ionic equilibria.

The student is assigned to a desk in the laboratory, in which he keeps his own individual set of apparatus, and is required to keep an accurate record of his experimental work in the form of a notebook. This notebook is periodically inspected.

So many of the chemical changes of matter being of a more or less spectacular nature, the young student finds himself engaged in a line of work that, from the first holds his interest and attention in a manner that few other subjects can claim.

Text-book: "Elementary Chemistry," by McPherson and Henderson.

"Exercises in Chemistry" by same author.

Course II. Lectures, two hours per week; Laboratory, two hours per week. Required of all Juniors electing chemistry.

This course is an extension of Course I. The study of reactions is at all times more thorough-going than in the preceding course. Considerable attention is directed toward the physical side of chemical action, and, throughout the course, problems bearing on the different principles under discussion are assigned for solution.

This course, and the course in qualitative analysis, are beautifully supplemental, in that the interpretations for the actions underlying analysis are here exhaustively considered.

Text-book.: "General Chemistry for Colleges," by Alexander Smith; "A Laboratory Outline for College Chemistry," by the same author.

Course III. Qualitative Analysis. Lectures one hour per week; Laboratory, four hours per week. Required of all Juniors electing chemistry.

This is a course of instruction in the detection, qualitatively, of unknown substances. At first, the student is given simple unknowns containing one of two metals. He learns the group reactions, whereby he may separate the different metals into groups, and the separation of these into smaller groups and individuals. Proceeding further in the study, he next encounters the reactions by means of which the commoner acids may be detected. As his knowledge and skill advance, he is finally given more or less complex commercial products and naturally occurring substances, such as portland cement, limestone, phosphate-rock, iron ores, etc., for analysis. He is, at all times required to understand the reactions and principles underlying his work. In this manner a course in qualitative analysis becomes more than its name implies, since it also constitutes an excellent course in general chemistry at the same time.

Text-books: "Qualitative Chemical Analysis," by Olin Freeman Tower.

Course IV. Quantitative Analysis. Lectures, one hour per week; Laboratory, four hours per week. Required of all Seniors who elect chemistry, biology, or electrical courses.

This course supplements the preceding one in that the student is now taught to estimate substances quantitatively. He begins by acquiring skill in the carrying out of standard methods for the determination of the commonly occurring metals and acids. As his skill increases, he is given more and more complex materials, finally acquiring the ability to carry out the quantitative analysis of ordinary commercial and natural products. His training includes estimations both in a gravimetric as well as in volumetric manner. This course, together with the preceding one. constitutes excellent training for those cadets who anticipate entering the profession of pharmacists, or who aspire to enter chemical laboratories as assistants.

Text-book: "Quantitative Chemical Analysis," by Talbot.

Course V. Organic Chemistry. Lectures, two hours per week; Laboratory, two hours per week. Required of all Seniors who elect chemistry.

This course consists of the study of organic general reactions, as usual at the beginning of this branch of the science. The actions discussed in the classroom are amply illustrated in the laboratory by the student's individual work. Here, he undertakes the carrying out of simple syntheses and reactions to emphasize the theoretical principles discussed in the lecture-room. This course furnishes excellent preliminary training for those who propose to undertake the study of medicine or of pharmacy, or for those who propose to enter the technical field. Aside from its value for the sake of the knowledge to be acquired, the course takes high rank on a purely educational basis, since it requires a high degree of mentality for its accomplishment.

Text-book: "Theoretical Organic Chemistry," by Julius B. Cohen.

DEPARTMENT OF BIOLOGY.

Major Knox

This department consists of but one standard college course of two lectures and two laboratory hours per week, and is the usual elementary course offered in Zoology.

The course is designed primarily for the benefit of those students intending to take up the study of medicine, and with this in view it has been assigned to collaborate with the requirements for admission to the medical colleges. Whereas this is the primary purpose for which the course was instituted, there is no study that so teems with interest, and which offers greater cultural opportunities to the student.

In the lecture-room, examples from the various classes in the animal kingdom are critically studied, this study being supplemented by the microscopic examinations and the dissections undertaken by the student in the laboratory.

The course is required of all chemistry-biology electives, and is also open for election to the English electives.

Text-book: "Manual of Zoology," by Parker and Haswell; "Invertebrate Zoology," by G. A. Drew.

DEPARTMENT OF PHYSICS.

Major Ferguson
Lieut. Clarke.

The department of Physics consists of a general laboratory with ample floor space, two lecture rooms, a dark room and suitable apparatus and storage rooms. It is equipped with modern physical apparatus and conveniences for individual laboratory practice and lecture table experiments.

Physics I. Elementary College Physics. Two hours a week lecture and recitation, and two hours a week laboratory practice. (Required of all Freshmen).

This course assumes no previous knowledge of Physics and consists of lectures, demonstrations, and problems presented as simply and directly as possible. It covers the entire field of general physics, mechanics, heat, light, sound, and electricity, in so far as time permits. The greatest stress will be laid on the foundation principles that prepare for practical specialization in the scientific fields. In the laboratory the students learn to use scientific instruments and to work out scientific problems with accuracy and efficiency.

Text-book. Kimball's "College Physics."

Physics II. (Junior Class Elective). General Physics, covering some of the more difficult and mathematical phases of subject. Special attention will be given to physical principles and problems related to engineering subjects. Topics that have been omitted from Physics I will be taken up in this course, thus rounding out the student's knowledge of general physics.

Text-book. Duff's Physics.

Physics III.—(Senior Class Elective). Electricity and Magnetism, two hours per week lecture and recitation, two hours per week laboratory.

This course is designed to present a working knowledge of practical electrical problems such as are evolved in generation, storage, distribution and utilization of electric power and in the measurements of electrical quantities.

Physics IV. (Senior Class Elective). An advanced course in some phase of Advanced Physics which will be chosen with a view to the needs of students electing the course. A choice will be given of a study of the principles of thermodynamics, Wireless Telegraphy, Telephony, and Physical Optics.

DEPARTMENT OF DRAWING.

Major LeTellier.

The work of this department includes: (1) a general course in Mechanical Drawing and Descriptive Geometry, which is required of all cadets; and (2) more advanced technical courses, which are required of cadets who elect the Engineering Course in the Junior and Senior years. The object of the first is to train the students in the use of Drawing and practical, graphical language, and to give them the advantage of the rigorous training derived from a course in Descriptive Geometry; the object of the technical courses is to develop graphic methods of investigating and solving engineering problems.

The method of instruction is based on the assumption that real mastery of engineering subjects can be developed only by constantly working problems. In assigning problems, effort is made to relieve the student of the waste of time incident to transcribing data and diagrams, or to repeating well-understood or purely mechanical operations. The problem sheets are given out with all data in such form that the student can begin without delay work on the essential part of the problems. No mere copying exercises are given. The text-books are supplemented by frequent lectures, and notes prepared by the instructor and furnished to the students in mimeographed form.

The drawing room is located on the second floor of the King Street Building, and is large, well lighted, and fully equipped. The equipment includes the following articles: Forty drawing tables, twenty-five of which are new adjustable tables of the most modern design, stools for all tables, eight large locker cases, two filing cases for problem sheets and specimen drawings, adjustable blackboard, large blueprinting frame mounted on track to facilitate exposure, blueprint tubes, pantograph for accurate transcribing, polar planimeter, copying machine for preparing notes, a collection of mechanical, architectural, topographic, and structural drawings, and a collection of models and machine parts.

Course I. SOPHOMORE CLASS. Required of all cadets Four hours a week.

MECHANICAL DRAWING—Preliminary work, to develop skill in handling drawing instruments; practice plates; useful geometrical constructions; construction of the conic curves, cycloids, involutes, and spirals; Reinhardt's system of freehand lettering; elementary orthographic projection, the object of which is to prepare the student for the work in Descriptive Geometry which follows in the Second Term. Fifty hours, fourteen plates.

Text-book. French's Engineering Drawing.

DESCRIPTIVE GEOMETRY—Great importance is attached to this work, as Descriptive Geometry supplies the principles on which all geometrical representation is based, as well as affording the best possible means for cultivating the geometrical imagination.

The course includes a series of problems of gradually increasing difficulty. All problems are demonstrated by models constructed by the instructor, and following this each cadet is required to construct his own models until he has made sufficient progress to dispense with them. Thirty problems, twenty hours.

Text-book: Church' Descriptive Geometry, and notes prepared by the instructor.

APPLIED DESCRIPTIVE GEOMETRY—The foregoing course in Descriptive Geometry is followed by a course the aim of which is to gradually adapt the principles of the subject to the practical problems of drawing and design. The use of auxiliary planes of projection, sectional views, and revolved views, is explained and problems involving their use are solved. This is followed by a series of problems in the intersection and development of surfaces. These problems deal with the usual geometrical solids, transition pieces, connecting rods, and various articles constructed from sheet metal. Twenty-two problems, thirty hours.

Text-book: French's Engineering Drawing, and notes prepared by the instructor.

PICTORIAL REPRESENTATION—Isometric, oblique, cabinet, diametric, and clinographic projection. The training in orthographic projection enables the student to master the essentials of these methods of representation in a short time. Twelve problems, twenty hours. Same text as above.

Course II. JUNIOR CLASS. Engineering Elective. Two hours a week.

MECHANICAL DRAWING—During the first term of this year, the work of the Sophomore Class in Mechanical Drawing is continued, the object being to give as much training in this subject as may be regarded as a safe minimum for students of civil engineering. The course begins with a brief study of the more important standard machine parts, such as bolts, nuts, screw-thread, springs, structural shapes, and rivets; this includes the correct and conventional methods of representation. Examination of government and manufacturer's drawings, with exercises based on these drawings; detailed and assembly drawings of simple machines and machine parts; tracing and blueprinting; drawing office system. Forty hours.

Text-book: French's Engineering Drawing.

ELEMENTARY KINEMATICS—Problems in the design of cams, quick-return motions, engine movements, valve gears. Fifteen problems, fifteen hours.

Text-book: Barr's Kinematics of Mechanism.

TOPOGRAPHICAL DRAWING—Plotting angles, plotting surveyors' notes, topographical symbols, contour mapping, including problems in grading, visibility, and map scales. Sixteen hours, ten problems.

Text-book: Raymond's Surveying.

Course III. SENIOR CLASS. Engineering Elective. Four hours a week.

EARTHWORK COMPUTATIONS—Use of contour maps in engineering operations; derivation of prismodical, prismatic, and approximate formulae for earthwork computations, and the application of these formulate to problems involving fills, excavations, grading, volume of water impounded by dams, and similar engineering operations. Twenty problems, including derivation of all formulae given in text. Twenty hours.

Text-book: Raymond's Surveying.

GRAPHIC STATICS—The object of this course is to give a thorough working knowledge of the principles of graphic statics, and to develop mathematical proofs for all of the methods employed. Graphic statics as a method of analysis is an essential part of an engineer's training, and, in order to master the subject, it is necessary to establish the mathematical soundness as well as the practical efficiency of its methods of analysis. The method of instruction is as follows: A set of problems involving the computation of reactions and stress in roof trusses are given. The graphical methods are given to the students without proofs, and they are shown how to apply the methods to the solution of the problems. In this way the efficiency of the methods is brought out, and the student's natural interest to find why these methods afford such easy solutions for intricate problems, is aroused. Then the subject is taken up with the simplest diagrams, and the proofs are gradually developed by a re-examination of the problems previously solved.

The order of the work is as fololws: General principles of graphic statics, composition, resolution, and equilibrium of forces, the force and funicular polygon, graphic moments, center of gravity, moment of inertia. Dead and wind loads on structures, Duchemin's formula for wind pressures, wall reactions, types of roof trusses, stresses in cantilever and unsymmetrical trusses, counterbracing, three-hinged arches. Graphic treatment of loads in beams. Types of bridges, train loads, Cooper's tables of train loads, analysis of loads in bridges. Fifty-eight problems, sixty hours.

Text-book: Malcolm's Graphic Statics.

STRUCTURAL DRAWING AND ELEMENTARY DESIGN—The object of this course is to supplement the course in Graphic Statics with an elementary knowledge of the materials and methods employed in structures. A study of standard structural shapes is made from the catalogs of the steel manufacturers; this is followed by a study of standard joints and connections, methods of designing and fabricating steel structures, and a complete study of a steel plate girder and a steel highway bridge. Twenty problems, forty hours.

*Text-book*s Conklin's Structural Steel Drafting and Elementary Design, Cambria Steel Company Handbook.

DEPARTMENT OF MODERN LANGUAGES.

All cadets are required to take French in the Freshman and Sophomore Classes. In the Junior Class, the student may elect (1) to continue the study of French in French III and French IV, (2) to begin and continue the study of German in German I and German II, (3) to begin and continue the study of Spanish in Spanish I and Spanish II.

N. B.—None of these elective courses will be offered unless the number of students applying warrants the formation of a section.

FRENCH.

LIEUT. VURPILLOT

Course I. FRESHMAN CLASS.—Three hours per week. Required of all.

This course is offered for beginners in French. The elements of French Grammar are studied, and especial attention is given to smooth translation, practice in grammatical forms, pronunciation, and the writing of French from dictation. This year the class is using Fraser and Squair's Elementary French Grammar for the study of grammatical forms, the writing of exercises, and the reading of about 50 pages of French prose and poetry.

MAJOR GRAESER.

Course II. SOPHOMORE CLASS—Three hours per week. Required of all.

The reading matter of this course is found in "Features of French Life," Parts I and II.

If time permits an additional text will be used.

Syntactical facts are continuously impressed, as suggested from reading assignments, and irregular verbs are thoroughly learned. Unrelenting effort is exerted to insist upon an intelligible pronunciation.

Composition, sight-reading, dictation and simple conversation are constantly practiced, using Francois's "French Prose Composition" as a text.

MAJOR GRAESER.

Course III. JUNIOR CLASS—Three hours per week. Elective.

For the period of the war, and in view of our cadets' prospective service in France as officers, the character of this course has been entirely changed from one of literature to one of arms. The course in reading consists of rather long assignments in Erckmann-Chatrian's "Conscrit de 1813' (Holt's edition of 240 pages), which are paraphrased by the students in their own words in French and general outlines of the story as it progresses are required to be handed in in French from time to time.

Familiarity with conditions and terms of modern French army life is gained by the study of "Le Soldat Americain en France."

Levi's "French Composition" is used as a text and every opportunity for conversation on the part of the student is utilized to the utmost.

MAJOR GRAESER.

Course IV. SENIOR CLASS—Three hours per week. Special elective course for period of the war.

The reading matter in this course is made up of selections from well-known French books that have appeared during the war, such as: Goffic's "Dixmude", Barbusse's "Sous le Feu", Bordeau's "Verdun", Barres's "Dans un Jardin de Lorraine", etc.

In addition to formal composition and conversation as presented in Levi's "French Composition", assignments from French newspapers are studied and reported on.

GERMAN.

Major Graeser.

Course I. Junior Class—Three hours a week. Elective

In entering upon the subject of German, a minimum of grammar and a maximum of practice, as presented in Gohdes and Buschek's "Sprach-und Lesebuch," are furnished and the student begins at the earliest possible moment the reading of a course containing two hundred to two hundred and fifty pages from such texts as: Marchen und Erzahlungen, Gluck Auf, Herein, Gruss aus Deutschland, etc., that furnish the German viewpoint of army life and events, places of interest, historical and imposing personalities of earlier and more recent times. Pronunciation is improved by constant drill, and by writing from dictation.

German I may be substituted for French III.

Major Graeser.

Course II. Senior Class—Three hours per week.

This course comprises about five hundred pages of reading. For the cultural element, Bernhardt's Literaturgeschichte is studied, and Lessing's Minna or Nathan, Schiller's Ballads and Tell, and Goethe's Mesiterwerke (Bernhardt) are translated entirely or in part.

Much parallel reading is assigned.

SPANISH.

Major Graeser.

Course I. Junior Class—(Senior Class for 1919 only). Three hours per week. Elective.

Since the beginning of the war, the exclusive interest in French and Spanish has practically eliminated the study of German.

War conditions have stressed the importance of the vast commercial field that lies open to the enterprise of United States' business in the Spanish-speaking countries of South America. A knowledge of Spanish as given in the course outlined below should prove an asset of prime value to young men in many lines of commercial activity.

In this course Hills and Ford's "First Spanish Course" is used for the elements of grammar. Reading, vocabulary and simple conversation are made possible at the very beginning of this course by the use of Worman's "First Spanish Book."

Conversation is further developed by the use of Rosenthal's Linguistic Method and phonographic records.

Fuentes and Francois's "A Trip to Latin America" furnishes valuable geographic and ethnic information and additional practice in composition.

MAJOR GRAESER.

Course II. SENIOR CLASS—Three hours per week.

The work in Spanish I will be further developed by the study of syntax, constant drill on the Spanish verb, pronunciation, etc.

The text for reading will be some modern novel such as Valera's "Pepita Jiminez" or Galdos's "Dona Perfecta" and the Boletin de la Union Panamericana.

Formal composition will be supplied by such texts as Waxman's, Umphrey's or Broomhall's and commercial correspondence by one of the various texts on this subject.

DEPARTMENT OF MILITARY SCIENCE:

COL. STOGSDALL.
MAJ. LE TELLIER

The theoretical and practical instruction in this course is laid down by the War Department in the same regulation that govern the Senior Division of the Reserve Officers' Training Corps. Any cadet who desires may receive the generous provision of this law, provided he will spend part of his last two summers at The Citadel in camp, where he will have transportation, board, and clothing provided by the Government. In addition thereto, provided his work is satisfactory, he will receive an allowance of Government uniforms and military equipment for the entire college course; and for the Junior and Senior, an allowance of money from the Government that will pay his board.

It should be clearly understood that the cadet in no way obligates himself for any service during war, or any training or duties after graduation, by joining the Reserve Officers' Training Corps. This is a Training Corps only. There is no obligation involved, except to train while the individual remains a cadet. And this condition is terminated when the cadet ceases to be a member of the Corps, either by graduation or otherwise.

The work throughout the course is progressive in character. The Freshman, besides routine preliminary instruction in drill and calisthenics, is taught, among other things, how to shoot, how to draw a map, the value of personal hygiene, the military history of the United States, and the military obligations of citizenship.

The Sophomore, in addition to other work, is instructed in the refinements of shooting, in map reading, camp sanitation, and camp expedients. He is taught signalling, semaphore, and flag; how to construct to scale field works and bridges. He is instructed in first aid, and in the principles of patroling.

The Junior is given the practical duties of drilling and instructing others. In his Freshman and Sophomore years he drills as a subordinate; now he exercises command. He is instructed in military sketching; in problems involving the principles of the Art and Science of War; in the elements of international law; and in practical military engineering.

The Senior is intrusted with the most responsible military commands in the Corps. He is given practical military engineering; tactical problems; court-martial proceedings; the international relations of America from discovery to the present day; in the gradual growth of principles of International Law embodied in American diplomacy, legislation, and treaties; the Psychology of War; and the general principles of strategy, planned to show the intimate relationship between the statesman and the soldier.

It is planned for the two upper classes to go every spring to Fort Moultrie, a regular Army Coast Artillery Post, situated in Charleston Harbor, to receive theoretical and practical instruction in handling seacoast guns and mortars.

The Citadel is not only being yearly rated as distinguished by the War Department, but is doing much more work than is required. The military training includes in its course so much field work and life in camp, target practice, and the practical solution of tactical problems in the field, that the cadet, provided his work has been thoroughly satisfactory, will be enabled upon its completion to fulfill in practice the obligations of merit and honor expected from graduates of this institution.

Capt. Le Tellier, assistant in the Department of Mliitary Science and Tactics, has been placed in charge of the instruction in Military Map making and Military Field Engineering.

DEGREES.

Upon the completion of the four years' course of studies, as outlined in the preceding pages, the cadet is granted the degree of Bachelor of Science.

The degree of civil engineer is granted to those students in Engineering who, after graduation, furnish satisfactory evidence of engineering work of a superior quality extending over at least three years, and who submit a satisfactory thesis.

BENEFICIARY SCHOLARSHIPS.

The State of South Carolina appropriates annually twenty thousand dollars for the suport of sixty-eight Cadets in the College. These scholarships are distributed among the various counties, as shown on pages 71-73.

Notices of vacancies in these scholarships are advertised in the month of July each year in the newspapers of the counties where they exist, and also in the leading State daily papers. Applications for these beneficiary appointments must be made upon printed forms furnished by the superintendent of The Citadel, and must be filled out in every particular, and returned to the Superintendent at The Citadel, who will then lay them before the Board of Visitors for their approval.

As it is the intention of the State to limit the beneficiary appointments to worthy young men without means of obtaining a college education, certificates of inability to pay are required in these applications.

The following are not eligible for beneficiary scholarship appointments:

(a) A person who during the current year has won or holds a scholarship at another State institution.

(b) A person who has been in attendance at The Citadel or "any other institution for higher learning known as a College or University," provided, however, that this condition shall not apply if there are no other applicants for the scholarship.

(c) A person who has forfeited a scholarship at The Citadel or any other State institution by failure to maintain himself.

Applicants to be eligible must be not less than sixteen years of age nor more than twenty years of age on September 20. They must be at least five feet in height, physically able to do military duty, of good moral character, and must show in their certificates that they are financially unable to go to college at their own or parents' expense.

In all counties where vacancies occur, competitive examinations will be held on the second Friday in August, by the County Superintendent of Education, and awards will be made to the applicant making the best grades, if they are otherwise eligible.

CADETS HOLDING STATE BENEFICIARY SCHOLARSHIPS, 1918-1919.

*ABBEVILLE—G. T. Hagan, '19; J. W. Wilson, '19.
AIKEN—M. Surasky, '19; P. J. McLean, '21.
ANDERSON—J. L. Whitten, '20; H. N. Heckle, '22; F. R. McAlister, '22.
BAMBERG—R. C. Roberts, '21.
BARNWELL—J. J. Still, '19; C. P. Hayes, '21.

BEAUFORT—E. Adams, '22 (special award 1 year).
BERKLEY—V. Harvey, '21.
CALHOUN—S. B. Antley, '21.
CHARLESTON—J. T. Witsell, '19; W. C. Lucas, '21; J. A. Tiedemann, '21; H. W. Crouch, '22.
CHEROKEE—W. W. Tolleson, '22.
CHESTER—T. C. Latimer, '20.
CHESTERFIELD—J. H. Rivers, '19.
CLARENDON—W. C. Wolfe, '19.
COLLETON—H. C. Jones, '19.
DARLINGTON—W. E. Williams, '21; M. M. Harrall, '22, 1 year.
DILLON—E. E. Brown, '22 (special award 1 year).
DORCHESTER—E. W. Felder, '22.
EDGEFIELD—J. B. Hart, '19.
FAIRFIELD—W. O. Brice, '21.
FLORENCE—T. W. Ross, '21; one vacancy.
GEORGETOWN—F. W. Ford, '19.
GREENVILLE—T. T. Dill, '19; T. C. Cannon, '19; T. M. Mayfield. '22.
GREENWOOD—J. K. Coleman, '19; D. B. Alexander, '22.
HAMPTON—O. R. Moore, '22 (special award 1 year).
HORRY—J. P. Cartrette, '21.
JASPER—W. M. Smith, '21.
KERSHAW—G. W. Nicholson, '19.
LANCASTER—M. Poliakoff, '22.
LAURENS—J. D. Fuller, '19; Q. D. Gasque, '22.
LEE—W. B. Smith, '21.
LEXINGTON—J. C. Gall, '22.
MCCORMICK—(See Abbeville Note).
MARION—J. L. Platt, '21.
MARLBORO—W. G. Gibson, '21.
NEWBERRY—W. L. Hardemann, '22; P. L. Langford, '22.
OCONEE—B. N. Singleton, '21.
ORANGEBURG—M. K. Jeffords, '19; G. C. Wise, '21 W. P. Davis, '22.
PICKENS—J. E. Rogers, '22 (special award 1 year).
RICHLAND—F. A. Thompson, '19; S. M. Roper, '22.
SALUDA—J. Y. Turner, '22 (special award 1 year).
SPARTANBURG—R. S. Baynard, '20; I. M. Coleman, '21; L. E. Diltz, '22; J. P. Thomas, '20 (special award 1 year).
SUMTER—J. H. Sanders, '19; H. V. Bradley, '20.
UNION—J. R. Lawson, '19.

WILLIAMSBURG—E. C. Perry, '21; A. C. Wilkins, '21.
YORK—E. B. Glenn, '22; W. H. McCorkle, '22.

* One vacancy from Abbeville County, when it occurs, goes to McCormick County.

CADETS HOLDING CHARLESTON CITY SCHOLARSHIPS, 1918-1919.

A. S. REYNOLDS, '19.
S. WARLEY, '19.
P. C. DOYLE, '20.
W. A. DOTTERER, '21.
C. B. PRENTISS, '22.
J. B. WESTON, '22.

PART VI

GRADUATES OF THE CITADEL
1846-1918

NOTE.—Any person knowing of errors in the following register of graduates, will confer a favor by sending corrections and information to the Superintendent. Graduates whose names are in *black letters* are dead.

Class of 1846   Remarks

1—C. C. Tew, Founder and Principal Hillsboro Military Academy, Col. N. C. Troops, C. S. A.; killed at battle Sharpsburg.
2—R. G. White, Physician; Major Tenth S. C., C. S. A.
3—C. O. Lamotte, Lawyer; Captain P. A. C. S.
4—John L. Branch, Civil Engineer; Colonel First S. C. M., C. S. A.
5—W. J. Magill, Prof. Math. Georgia Mil. Inst.; Col. 1st Ga. Reg. C. S. A.
6—John H. Swift, Civil Engineer.

Class of 1847.

7—Johnson Hagood, Brig. Gen., C. S. A.; Governor of South Carolina.
8—E. L. Heriot, Civil Engineer.
9—S. B. Jones D.D., Minister; President Columbia College.
10—J. P. Southern, Banker.

Class of 1848

11—H. Oliver, Civil Engineer.
12—A. Buist, Captain S. C. V., C. S. A.; Minister.
13—J. W. Gregory, Captain Engineers, C. S. A.; Planter.
14—H. D. Kennedy, Professor History, Arsenal Academy.
15—F. F. Warley, Lawyer; Major Second Regiment S. C. Art.; C. S. A.
16—H. L. Brantley, Civil Engineer.
17—J. J. Matthews, Major Georgia Volunteers, S. C. A.
18—A. J. Jamison, Volunteer Service C. S. A.
19—J. D. Powell, Captain P. A. C. S.

Class of 1849

20—P. F. Stevens, Supt. Citadel, '59-'61; Col. Holcomb Legion, C. S. A.; Bishop Reformed Episcopal Church.
21—U. A. Rice, Captain 48th Ga. Vol., C. S. A. Physician; Georgia.
22—J. T. Zealy, Minister; President Winona Female Institute; Mississippi.
23—H. L. Thurston, Lawyer.
24—J. B. White, Superintendent Citadel Academy, 1861-1865.
25—G. B. Lartigue, Physician; Major on Gen. Hagood's Staff, C. S. A.
26—W. G. Inglesby, Physician.

Class of 1849   Remarks

27—G. H. Bunker, Civil Engineer.
28—J. A. Walker, Minister, Texas.
29—H. W. Stewart, Engineer Corps, C. S. A.
30—T. E. Strother, Lieutenant City Guard, Charleston.
31—W. M. Morgan, Bookkeeper.

Class of 1850

32—D. H. Eggleston, Professor, Mount Zion College, Winnsboro.
33—J. A. Houser, Captain Ga. Vol.; C. S. A.; Planter, Georgia.
34—C. D. Oliver, Civil Engineer.
35—J. W. Robertson, Col. 37th Ala., C. S. A.; Pres. Roswell Mfg. Co.; Adj. Gen. State of Georgia.
36—J. A. Crooker, Civil Engineer; Lieutenant 27th S. C. V., C. S. A.
37—O. A. Darby, D.D., Minister; President Columbia Female College.
38—S. N. Kennerly, Physician; First Lt. 25th S. C. Regt. C. S. A.
39—J. R. Abrams, Civil Engineer, Chattanooga, Tenn.
40—J. L. Inglesby, Insurance Office, Charleston.
41—G. L. Odom, Physician.
42—H. N. S. Wheaton, Lawyer; Volunteer Service, C. S. A. Texas.
43—A. L. Edwards, Planter.

Class of 1851

44—J. P. Thomas, Founder and Supt. Carolina Mil. Inst., 1873-1882; Supt. Citadel, 1882-1885.
45—W. H. Wright, Physician.
46—J. G. Pressley, Lawyer; Lt. Col. 25th S. C. V., C. S. A.; Judge of Superior Court, California.
47—W. W. Veitch, Physician.
48—N. W. Armstrong, Professor of Mathematics Citadel Academy, 1859...
49—L. A. Brown, Civil Engineer.
50—J. B. Chandler, Planter; Maj. Reg. S. C. Reserves.
51—J. M. Pelot, Physician, Fifth Regiment, S. C., C. S. A.
52—J. J. Lucas Major Lucas' Battalion Heavy Art., Regulars C. S. A.; Planter; Director A. C. L. R. R.; Member Board of Visitors The Citadel.
53—James Aiken, Lawyer; Lt.-Col. 13th Ala., C. S. A.; Judge Supreme Court, Alabama.
54—J. W. Hudson, Physician; Assistant Surgeon 4th S. C., C. S. A.
55—B. W. Powell, Captain Fla. Vol., C. S. A., Merchant, Florida.
56—E. J. Walker, Lawyer; Colonel Georgia Volunteers, C. S. A.
57—T. J. Arnold Civil Engineer.
58—J. B. Cottrell, D.D., Minister; Captain Alabama Vol., C. S. A.
59—W. S. Dudley, Physician.
60—E. J. Frederick, Physician; Adjutant Lamar's Art., C. S. A.
61—E. C. Bailey, Planter.
62—J. L. Seabrook, Planter; Captain Third Regt. S. C. Cav., C. S. A.
63—J. B. Colding, Lawyer; Capt. Ga. Vol., C. S. A. killed at Winchester, 1863.
64—H. S. Bass, Captain City Guards, Charleston.

Class of 1851

65—F. G. Palmer, Civil Eigineer; Lt.-Col. Holcombe Legion, C. S. A.; Mortally wounded at Second Manassas.
66—W. R. Powell, Civil Engineer; Captain 2d S. C., C. S. A., California.
67—T. H. Cook, Lawyer; Lieutenant First S. C., C. S. A.
68—S. Collins, Planter.
69—W. D. McMillan, Captain 11th Regt., S. C. V., C. S. A.; Minister.

Class of 1852

70—A. H. Little, Veteran Mexican War; died 1854.
71—D. T. Williams, Lawyer; Killed in Battle Gettysburg.
72—W. S. Brewster Lawyer; Capt. Ga. Vol. C. S. A.; Killed in Battle Fredericksburg.
73—G. W. Earle, Civil Engineer; Captain Artillery, C. S. A.
74—C. S. Gadsden, Maj. 1st S. C., C. S. A.; Pres. N. E. R. R.; Chairman Board of Visitors The Citadel; Cherleston, S. C.
75—W. Y. McCammon, Principal Military Academy, Alabama.
76—W. H. Dial, Captain Florida Volunteers, C. S. A.; Merchant.
77—T. W. Fitzgerald, Teacher; Capt. 12th Ala. Regt., C. S. A.; Mortally wounded at Chancellorsville.
78—J. W. Murray Minister.
79—S. C. DePass, Adjutant First Ga. Regt., C. S. A.; Cotton Buyer.
80—R. A. Palmer, Lt. Miss. Vol., C. S. A.; Killed at First Manassas.
81—H. B. Houseal, Lt. Co. H., 7th Fla., Vol., C. S. A.; Died in service, 1862.
82—G. W. Seabrook, Planter; Died 1862.
83—C. S. Henagan, Teacher.
84—J. W. Daniels, Teacher; Captain Palmetto Sharpshooters, C. S. A.
85—G. E. Gamble, Planter; Died in service, C. S. A.
86—John C. Rich, Physician.
87—P. A. Raysor, Planter; Captain Cavalry, C. S. A.
88—S. M. J. Prothro Physician; Captain Georgia Vol., C. S. A.

Class of 1854

89—M. Jenkins, Prin. Yorkville Mil. Acad.; Brig.-Gen. C. S. A.; Killed at battle of Wilderness.
90—Thomas E. Hart, Ph.D., Heidelberg; Prof. Chemistry, Furman University.
91—A. D. Hoke, Physician; Captain Second S. C., C. S. A.
92—J. J. Jenkins, Died 1855.
93—A. Coward, Col. 5th S. C., C. S. A.; Supt. K. M. M. S.; Supt. Citadel, 1890-1908; Orangeburg, S. C.
94—J. D. Radcliffe, Colonel 18th N. C., C. S. A.; Merchant, Augusta, Ga.
95—C. T. Haskell, Civil Engineer; Captain First South Carolina, C. S. A.; Killed in battle on Morris Island 1863.
96—Cicero Adams, Lawyer; Major Twenty-second S. C., C. S. A.
97—J. M. Steadman, Merchant; Colonel Sixth S. C., C. S. A.
98—D. G. Fleming, Civil Engineer; Capt. S. C. Art., C. S. A.; Killed at explosion of mine near Petersburg.
99—A. H. Mazyck, Lt. Battalion State Cadets; Bookkeeper, Charleston, S. C.

Class of 1854   Remarks

100—J. F. Culpepper, Physician; Capt. Palmetto Batt., C. S. A.; Timmonsville, S. C.
101—D. R. Jamison, Lawyer; Aide to General Jenkins, C. S. A.

Class of 1855

102—W. P. DuBose, Adj. Holcombe Legion, C. S. A.; Prof. Univ. of the South, Sewanee, Tenn.
103—John D. Wylie, Lawyer; Lieutenant-Colonel S. C., C. S. A.
104—P. Bryce Physician, General Morgan's Staff, C. S. A.
105—J. B. Patrick, Lt. Battalion State Cadets; Founder Patrick Mil. Inst.
106—W. F. Nance, Major and A. A.-Gen. A. N. V., C. S. A.
107—B. Burg Smith, Col. 16th and 24th Regt., S. C., C. S. A.; Civil Engineer in charge 6th Lighthouse District.
108—W. D. Gaillard, Professor Hillsboro Military Acad.; Died 1860.
109—J. E. Presley, Physician; Colonel Tenth S. C., C. S. A.
110—Thomas E. Lucas, Physician; Maj. 8th S. C., C. S. A., Chesterfield, S. C.
111—P. S. Kirk, Physician; Surg. Longstreet's Corps, C. S. A. Trial S. C.
112—W. J. Crawley, Teacher; Lt.-Col. Holcombe Legion, C. S. A.
113—F. L. Parker, Chief Surgeon, Maj.-Gen. Field's Div., Longstreet's Corps. C. S. A.; Dean Medical College of S. C., Charleston, S. C.
114—R. C. Carlisle, Physician; Asst. Surgeon, P. A. C. S.
115—J. S. Mixon, Planter; Lieutenant Hagood's Regiment, C. S. A.
116—J. M. Dean, Planter; Lt.-Col. 7th Regt. Ark. Vol., C. S. A.; Killed at Battle Shiloh.
117—E. White, Civil Engineer; Assistant Engineer P. A. C. S.
118—J. Venning, Planter; Lientenant in White's Bat. Ark., C. S. A.

Class of 1856

119—J. F. Lanneau, Capt. Cav. Hampton Legion, C. S. A.; Prof. Math., Wake Forest, N. C.
120—W. R. Erwin Merchant; Died 1857.
121—I. G. W. Steadman, Retired Physician and Manufacturer; Col. 1st Ala. Volunteers, C. S. A.; St. Louis, Mo.
122—E. M. Law, Major-General A. N. V.; Superintendent Military School; Bartow, Fla.
123—E. Croft, Lieutenant-Colonel 14th S. C. Vol., C. S. A.; Lawyer.
124—H. S. Thompson, Prof. French Arsenal Acad.; Governor of South Carolina; Asst. Sec. of Treas.; Comp. N. Y. Life Ins. Co.
125—J. D. Nance, Lawyer; Col. 3d S. C. Vol., C. S. A.; Killed in battle of Wilderness.
126—J. A. Evans, Killed in Battle of Kennesaw Mountain, 1864.
127—G. Ross, Physician; Captain Arkansas Volunteers, C. S. A.
128—L. F. Dozier, Physician; Asst. Surgeon Longstreet's Corps; Anderson, Cal.
129—R. M. Sims, Planter; Adj. and Ins.-Gen. Longstreet's Staff, C. S. A.; Sec. of State of South Carolina.
130—R. Y. Dwight Physician; Assistant Surgeon P. A. S., Pinopolis, S. C.
131—A. M. McAlister, Teacher, Alabama.
132—J. A. Finch, Merchant; Vol. 6th S. C.; Killed in 2d Battle Manassas.
133—A. Y. Lee, Architect; Lieutenant Artillery, C. S. A.

Class of 1857   Remarks

134—W. M. Tennent, Lawyer; Captain Engineering Corps, C. S. A.
135—V. E. Manget, Professor in Georgia Female College; Capt. Bn. Ga. Cadets.
136—R. K. Thomas, Professor King's Mt. Mil. School; Died 1860.
137—W. J. Davis, Capt. 1st Regt., S. C. Inf., C. S. A.; Editor and Lawyer, Louisville, Ky.
138—J. E. Black, Adjutant P. A. C. S.; Insurance Agent, Arkansas.
139—H. B. D'Oyley, Teacher; Died 1859.
140—H. D. Moore, D.D., Chaplain 12th Ala., C. S. A.; Pres. Ala. College.
141—T. S. Hemingway, Physician; Assistant Surgeon, P. A. C. S.
142—J. M. Adams, Teacher; Maj. and Brig. Q. M. S. C. Vol., C. S. A.
143—B. M. Walpole, Lieutenant Volunteers, C. S. A.
144—H. A. Gaillard, Lawyer; Maj. 6th S. C., C. S. A.; Planter, Winnsboro, S. C.
145—T. H. Mangum, Maj. C. S. A., Commanding Post Meridian, Miss.; Physician, Trenton, Texas.
146—C. W. McCreary, Teacher; Colonel 1st S. C., C. S. A.; Killed at Five Forks, Va.
147—R. T. Harper, Civil Engineer; Lt. Eng. Corps Hood's Div. C. S. A., Gastonia, S. C.
148—J. K. Garmany, Volunteer service, C. S. Navy; Merchant.
149—W. Z. Bedon, Physician; Surgeon P. A. C. S.
150—J. F. Hart, Lawyer; Major Horse Artillery Battalion, A. N. V.
151—H. D. Garden, Lawyer; Capt. and Ins.-Gen. Gregg's Staff, C. S. A.
152—R. Campbell, Lawyer; Lieutenant-Colonel 11th S. C., C. S. A.
153—Ellison Capers, Brigadier-General, C. S. A.; Bishop P. E. Church.

NOTE:—The time of the Annual Commencement having been changed from November to April, there were no graduates for 1858.

Class of 1859

154—T. H. Law, Minister, Spartanburg, S. C.
155—P. S. Layton, Teacher, Colonel Fourth Miss. Regt., C. S. A.
156—W. P. Shooter, Lawyer; Lt. Col. 1st S. C., C. S. A., Killed in Battle in Va., 1864.
157—Warren Adams, Prof. Hillsboro Mil. Acad.; Lt.-Col. 1st S. C. Regt., C. S. A.
158—T. A. Huguenin, Major First S. C., C. S. A.
159—J. L. Litchfield, Lawyer; Capt. 7th S. C.; C.| S. A.; Mortally wounded in Battle of Maryland Heights, 1862.
160—O. J. Youmans, Lawyer; Col Second S. C. Vol., C. S. A.; Mortally wounded in Battle near Richmond, 1864.
161—W. E. Cothran, Planter; Captain Seventh S. C. C. S. A.
162—G. M. McDowell, Merchant; Lt. S. C. V., C. S. A.; Killed at Gettysburg.
163—T. J. Weatherly, Physician; Asst. Surgeon 6th Ala., C. S. A. Dillon, S. C.
164—R. Press Smith, Physician; Maj. 1st S. C., C. S. A. Santa Rosa, Cal.

Class of 1859   Remarks

165—W. R. Marshall, Capt. Art., Army of the West, C. S. A.; Federal Civil Service.
166—T. O. McCaslan, Teacher; Vol. Services, C. S. A.; Killed in Battle in Virginia, 1862.
167—J. E. Spears, Lawyer; Captain Twenty-fourth S. C., C. S. A.
168—F. L. Garvin, Captain Palmetto Sharpshooters, A. N. V.

Class of 1860

169—F. H. Harleston, Civil Engineer; Capt. Art., C. S. A.; Killed at Fort Sumter, 1863.
170—A. J. Norris, Lawyer; Capt. Lucas' Batt. Heavy Art. Regulars, C. S. A.
171—A. S. Gaillard, Prof. Hillsboro Mil. Acad.; Capt. C. S. A.; Died in 1870 of wounds received in service.
172—W. E. Stoney, Capt. on Gen. Hagood's Staff, C. S. A.; Comptroller-Gen. S. C.
173—E. A. Erwin, Lt. 1st S. C., C. S. A.; Killed at siege of Charleston, 1863.
174—S. S. Kirby, Lt. Palmetto Batt. Art., C. S. A.; Killed at River Bridge, S. C., 1865.
175—F. DeCaradeuc, Scout Army N. Va.; Wounded; died 1862.

Class of 1861

176—C. I. Walker, Lt.-Col. Tenth S. C., C. S. A. Charleston.
177—J. D. Lee, Adjutant Palmetto Sharpshooters C. S. A.; Killed at Battle of Frazier's Farm, 1862.
178—J. A. Tennant, Adj. Third N. C., C. S. A.; Architect, Asheville, N. C.
179—T. G. Dargan, Lieutenant Artillery, C. S. A.
180—R. O. Sams, Prof. Math. Montgomery Mil. Acad.; Teacher, Jonesville, S. C.
181—S. B. Pickens, Colonel 12th Ala., C. S. A.; G. F. Agt. S. C. R. R.
182—J. H. Burns, Maj. Fifth N. C., C. S. A.; killed at Gettysburg.
183—J. M. Whilden, Maj. 23rd S. C., C. S. A.; killed at Second Manassas.
184—S. C. Boylston, Adj. 1st S. C. Art.; Manager Columbia (S. C.) Granite plant.
185—T. M. Wylie, Lt. Sixth S. C., C. S. A.; died of wounds 1865.
186—J. C. Palmer, Adjutant 24th S. C., C. S. A.; Killed at Chickamauga.
187—G. E. Haynesworth, Lieutenant Artillery, C. S. A.; Lawyer.
188—W. B. Guerard, Lieutenant Engineers, P. A. C. S.; Civil Engineer.
189—N. Wilson, Drillmaster, C. S. A.; Killed at Sharpsburg.
190—J. S. Austin, Capt. C. S. A.; Pres. Pacific Meth. College, Santa Rosa, Cal.
191—R. Croft, Lieutenant South Carolina Artillery, C. S. A.
192—T. E. Raysor, Captain Eleventh S. C., C. S. A.; Teacher
193—W. C. Vance, Volunter C. S. A.
194—J. L. S. Dove, First Lieutenant Palmetto Light Art., C. S. A.
195—Ralph Nesbit, Colonel C. S. A.; Rice Planter, Waverly Mills, S. C.
196—W. S. Simpkins, Lt. 1st S. C. Art., C. S. A.; Lawyer, Dallas, Texas.

Class of 1861   Remarks

197—J. A. Keith, Lieut. Lucas' Batt. Heavy Art., Regulars C. S. A.; Physician.
198—J. T. Morrison, Lieutenant Eleventh S. C., C. S. A.; Teacher.
199—C. H. Ragsdale, Lieutenant South Carolina Cavalry.
200—James Thurston, Lieutenant Marines South Carolina; Navy.
201—T. B. Ferguson, Union Club, New York, N. Y.

Class of 1862

202—George G. Wells, Prof. Hillsboro Mil. Acad.; Lawyer, Greenville, S. C.
203—Wm. F. McKewn, Prof. Math. Montgomery Mil. Acad., Ala.; Vol. 5th Regt.; Mortally wounded at Fredericksburg.
204—Amory Coffin Jr., First Sergeant Marion Artillery, C. S. A.
205—Wm. B. McKee, Lieut. Palmetto Batt. Art.; Asst. to Vice-Pres. Plant Railways.
206—R. F. Lawton, Adjutant Second Georgia Cav., C. S. A.; Banker.
207—G. A. McDowell, Vol. Aiken's Regt., S. C. Cav.; Killed on John's Island, 1864.
208—S. D. Steedman, Adj. First Ala., C. S. A.; Lawyer, Steedman, Texas.
209—I. H. Moses, Volunteer in Aiken's Regt., S. C. Cav., C. S. A.
210—D. P. Campbell, Volunteer 11th S. C., C. S. A.; Killed at Pocotaglio.
211—S. P. Smith, Captain Siege Train, S. C., C. S. A., Charleston S. C.
212—Wm. M. Tucker, Vol. Hampton's Legion, C. S. A.; Prof. Hillsboro Mil. Acad.
213—L. R. Stark, Adjutant Tenth S. C., C. S .A.; Physician in Arkansas.
214—J. R. Mew, Vol. S. C. Art., C. S. A.; Civil Engineer, C. & S. Railroad.
215—M. S. Elliot, Vol. S. C. Art., C. S. A., Planter, Beaufort, S. C.
216—J. L. Taylor, Drillmaster Twenty-second S. C., C. S. A.
217—Gerard B. Dyer, Vol. Second S. C. C. S. A.; Killed in battle near Richmond, 1864.
218—Wm. H. Brice, Vol. Service, N. C., C. S. A; Mercantile Business, Boston.
219—John B. Allison, Lieutenant Twelfth Georgia Art., C. S. A.

Class of 1863

220—M. M. Farrow, Prof. French, Hillsboro Mil. Acad.; Lt. Engineers. C. S. A.
221—R. H. Griffin, Lieutenant Pontoniers, A. N. V.; In business North.
222—J. K. Law, Aide to Gen. Law, C. S. A.; Judge of Superior Court, Mercer, Cal.
223—F. M. Farr, Captain Fifteenth S. C. C. S. A.; Banker, Union S. C.
224—B. G. Rushing, Teacher.
225—A. Doty, Signal Corps, C. S. A.; Teacher.
226—H. W. DeSaussure, Lieutenant First S. C. Art., C. S. A.; Physician.
227—W. F. Rice, Volunteer service; Merchant.
228—R. L. Cooper, Lieutenant First S. C. Art., C. S. A.; Lawyer.
229—W. M. Smith, Adj. 27th S. C., C. S. A.; Mortally wounded at Cold Harbor.

Class of 1863   Remarks

230—B. R. Snead, Died 1863.
231—J. B. Dotterer, Sergt-Maj. 24th S. C., C. S. A.; Mortally wounded at Resaca, 1864.

Class 1864

232—P. S. Norris, Assistant Professor Hillsboro Military Academy.
233—C. H. Rice, Volunteer Hart's Battery, A. N. V.
234—L. W. Kennedy, Volunteer 26th S. C., C. S. A., Farmer.
235—A. N. Alexander, Farmer.
236—N. W. Steedman, Volunteer 26th S. C., C. S. A.; Farmer in Texas.
237—J. V. Morrison, Lipscomb's Regt., C. S. A.; Farmer and Merchant, Hampton County.
238—J. D. Quattlebaum, Adj. Twenty-second S. C., C. S. A.; Killed in explosion of mine at Petersburg, 1864.
239—J. U. Matthews, Volunteer 26th S. C., C. S. A.; Teacher.
240—A. G. Howard, Merchant in Georgia.
241—W. H. Mew, Civil Engineer.
242—W. P. Baskin.
243—J. H. Bouknight, Farmer, Johnston.
244—G. R. Dean, Physician, Spartanburg.
245—A. B. DeSaussure.
246—O. D. East.
247—J. M. Gray.
248—S. F. Hollingsworth.
249—C. W. Horsey, Physician.
250—J. W. King, Physician, Florence, S. C.

Class of 1865

251—G. W. Klinck.
252—H. Perroneau.
253—J. M. Rogers, Merchant, Winston-Salem, N. C.
254—O. Sheppard, Lawyer; Chairman Board of Visitors The Citadel; Edgefield, S. C.
255—W. N. Snowden, Merchant.
256—Edward Thomas, Railroad Service. Ticket Agent, Hope, Ark.
257—S. E. White, Planter.
258—W. R. Vernon.

NOTE:—The Institution was closed by the results of the War, and remained closed until 1882. Consequently there were no Graduates from 1865 to 1886.

Class of 1886

259—R. M. Walker, Engineer and Contractor, Atlanta, Ga.
260—T. P. Harrison, Prof. English, A. & M. College, Raleigh, N. C.; Ph.D. Johns Hopkins University.
261—O. J. Bond, Superintendent The Citadel.
262—F. J. Devereaux.

Class 1886   Remarks

263—G. M. Gadsden, Civil Engineer, Savannah, Ga.
264—J. P. Kinard, Ph.D., Johns Hopkins University. Professor Winthrop College, Rock Hill, S. C.
265—A. J. Howard, Farmer, Darlington, S. C.
266—W. G. Jeffords, Charleston, S. C.
267—Edward Anderson, Capt. Heavy Art., S. C. Vol., U. S. A.; Real Estate Jacksonville, Fla.; U. S. R., Major, Staff.
268—H. C. Schirmer, Rice Merchant, Houston, Texas.
269—**Wm. Jennings.**
270—J. T. Coleman, Dist, Agt. Prudential Ins. Company, Charleston, S. C.
271—**S. C. Boylston, Jr.**
272—J. R. McCown, Teacher, Florence, S. C.
273—F. M. Robertson, Insurance, Charleston, S. C.
274—A. W. Lawton, Farmer, Lena, S. C.
275—J. W. Gibbes, Merchant, Columbia, S. C.
276—W. D. Gaillard, Fer. Company, Charleston, S. C.
277—P. N. Timmerman, Railroad Service, Florence, S. C.
278—C. L. Wroton, Traveling Salesman, Rock Hill, S. C.
279—Archie China, Physician, Sumter, S. C.
280—**W. G. Workman.**
281—**Benj. Munnerlyn.**
282—F. O. Spain, Agent D. C. Heath Company, Publishers, Gainesville, Fla.
283—**B. C. Jennings.**
284—R. T. Crawford, Civil Engineer, Charleston, S. C.
285—L. S. Carson, Lt.-Col., Inf., U. S. A.
286—W. L. Floyd, Prof. Botany and Horticulture, Univ. of Fla., Gainesville, Fla.
287—E. M. Law, Prof. Chem. and Phys. South Fla. Mil. Acad., Bartow, Fla.
288—R. B. Furman, Physician, Privateer, S. C.
289—S. R. Kirton, Civil Engineer, Homerville, Ga.
290—**W. B. Weathersbee.**
291—W. A. Leland, Civil Engineer, Johnson City, Tenn.
292—E. C. McCants, Superintendent City Schools, Anderson, S. C.
293—**E. C. Youmans.**
294—J. K. Brockman, Manager Title and Guaranty Co., Birmingham, Ala.
295—H. F. Rice, District Judge South Carolina, Aiken, S. C.
296—C. G. White, Dentist, Charleston, S. C.
297—J. M. Allen, Commercial Traveler, Goldsboro, N. C.
298—E. W. Bell, Vice-President Georgia State Savings Association, Savannah, Ga.
299—T. H. Goethe, United States Pension Official, Greensboro, N. C.
300—**J. W. Ouzts.**
301—E. L. Price, Bamberg, S. C.
302—R. T. Wylie, Physician.
303—H. S. Hartzog, 1518 Vernon Avenue, St. Louis, Mo.
304—T. M. McCutcheon, Physician, Alcolu, S. C.
305—T. G. McMichael, Lawyer, Charlotte, N. C.
306—C. S. Evans, Physician, Clio, S. C.
307—**J. H. Brooks.**

Class of 1886

308—J. S. Cureton.
309—**Horatio Lenoir.**
310—W. F. Roberts, Major C. A. C., U. S. N. G.
311—N. S. Harris, Atlanta, Ga.

Class of 1887

312—G. A. Lucas, Commercial Traveler, Augusta, Ga.
313—A. M. Kennedy, Merchant, Williston, S. C.
314—**C. B. Ashley.**
315—**E. A. Laird.**
316—W. S. Allen, Merchant, Charleston, S. C.
317—I. I. Bagnall, Manning, S. C.
318—W. L. Bond, Druggist, Fredericksburg, Va.
319—R. R. Jeter, Secretary Glenn-Lowry Man. Con. Whitmire, S. C.
320—**H. H. Brunson.**
321—E. C. Lee, Railway Ticket Office, Charleston, S. C.

Class of 1888

322—**B. L. Clark.**
323—M. W. Pyatt, Lawyer, Georgetown, S. C.
324—A. G. Miller, Superintendent Schools, Waycross, Ga.
325—F. H. Elmore, Southern Railway Official, Washington, D. C.
326—G. H. Cornelson, Minister.
327—J. H. Noland, Minister, S. C. Conference, M. E. Church, South.
328—A. N. Brunson, Minister, South Carolina Conference, M. E. Church, South; Member Board of Visitors The Citadel.
329—J. M. Patterson, Lawyer, Allendale, S. C.
330—J. R. Padgett, Merchandise Broker, Jacksonville, Fla.

Class of 1889

331—L. W. Haskell, United States Consul; Major N. G. S. C.
332—W. W. Lewis, Lawyer; Lieut.-Colonel U. S. N. G.
333—W. M. Smith, Civil Engineer, Barnes & Smith, Dayton, Ohio.
334—S. B. Platt, Superintendent Warwick Cotton Mills, Augusta, Ga.
335—M. L. Smith, Major, Judge Advocate, U. S. N. A.
336—C. E. Johnson, Teacher, Chicago, Ill.
337—W. C. Davis, Lawyer; Captain, U. S. V.; Manning, S. C.
338—**R. S. Clarkson.**
339—**W. H. Dial.**
340—R. B. Cunningham, Agnes Scott College, Decatur, Ga.
341—W. H. Rose, Secretary Cotton Mill, Columbia, S. C.
342—**D. McQ. Fraser.**
343—T. B. Haynesworth, Farmer, Florence, S. C.

Class of 1890

344—W. H. Simons, Colonel, U. S. A.
345—T. M. Hunter, Presbyterian Minister, Baton Rouge, La.
346—**J. E. Buzhardt.**

Class of 1890   Remarks

347—L. DeV. Blake, Secretary and Treasurer Cotton Mill, Belton, S. C.
348—L. L. Gaillard, Electrical Engineer, New England Eng. Co., Waterbury, Conn.
349—S. D. Lucas, Manager Southern Bell Telephone and Telegraph Co., Wilmington, N. C.
350—J. T. Boozer.
351—J. C. Bailey, Minister, Liberty, S. C.
352—John Ball, Vice-President and Manager Con. Gro. Co., Jacksonville, Fla.
353—E. C. Hughes, Asst. Secretary and Treasurer Union Naval Stores Co., Mobile, Ala.
354—A. G. Singletary, Insurance, New Roads, La.
355—G .W. Allison, Lawyer, San Francisco, Cal.
356—D. G. Dwight, Fertilizer Manufacturer, Charleston, S. C.
357—B. S. Cogburn, Teacher, Neeses, S. C.
358—William Godfrey, of Wm. Godfrey & Co., Cheraw, S. C.
359—A. L. Humphreys, Lawyer, Live Oak, Fla.
360—W. W. Dixon, Supt. Schools, St. Stephens, S. C.
361—W. E. Mikell, Dean of Law School, University of Pennsylvania. Philadelphia, Pa.
362—C. D. Gooch.
363—J. D. Nix, Lawyer, New Orleans, La., 400 Audubon, Bldg.
364—R. L. Dargan.
365—J. F. Evans, Real Estate, Anderson, S. C.
366—C. E. King, Physician, Mayesville, S. C.
367—P. B. Bird, United States Engineers, Jacksonville, Fla.
368—W. W. Tison, Physician, Cedartown, Ga.
369—F. C. Black, Supt. High School, Checatah, Okla.
370—E. R. Zemp, Physician, Knoxville, Tenn.
371—H. A. DeLorme, Physician, St. Louis, Mo.
372—W. W. Stewart.
373—J. G. Watts.
374—L. S. Trotti, Cashier Bank, Brookland, S. C.
375—F. M. Edwards, Civil Engineer, Jacksonville, Fla.
376—S. F. Garlington, Lawyer, Augusta, Ga.
377—F. B. Grier, Lawyer, Greenwood, S. C.
378—A. G. Guerard, of A. G. Guerard & Son, Home Insurance Company, Savannah, Ga.
379—Havelock Eaves, Major U. S. Volunteer; Cotton, Orangeburg, S. C.
380—J. T. Burdell, Civil Engineer, Tarboro, N. C.

Class of 1891

381—J. W. Perrin, General Freight Agent A. C. L., Wilmington, N. C.
382—T. J. Mauldin, Judge Thirteenth South Carolina Circuit, Pickens, S. C.
383—E. M. Whaley, Physician, Columbia, S. C.
384—H. W. Fraser, Cashier Bank, Georgetown, S. C.
385—D. D. Salley, Physician, Orangeburg, S. C.

Class of 1891   Remarks

386—D. A. Spivey, Cashier Bank, Conway, S. C.
387—E. M. Blythe, Lawyer, Greenville, S. C.; Former Col. 1st Regt., N. G., S. C.; Maj. Inf, N. A.
388—E. B. Lorick, Farmer, Camden, S. C.
389—R. C. Roberts, Dentist, Barnwell, S. C.
390—J. D. Frost, Captain and Adjutant U. S. V.; Maj. N. A.
391—W. N. Tillinghast, Minister, Church of the Epiphany, Washington, D. C.
392—A. F. Carter, Physician.
393—J. W. Magrath, Lawyer, 60 Wall Street, New York.
394—J. M. Robertson, President Porter-Snowden Company, Charleston, S. C.
395—W. C. Humphreys, Supt. Etiwan Fertilizer Company, Charleston, S. C.
396—P. K. McCully, Jr., Colonel U. S. N. G.
397—A. A. Aveilhe, with the Bartow Phosphate Company, Savannah, Ga.
398—J. L. Oliver.
399—A. M. Brailsford, Major U. S. N. G. Med. Corps.

Class of 1892

400—A. S. Thomas, Minister, P. E. Church, Columbia.
401—W. Z. McGhee.
402—G. R. Coffin, Lawyer, Augusta, Ga.
403—J. G. Beckwith.
404—A. G. Etheridge, Teacher, Texas.
405—J. F. McElwee, Merchant; York, S. C.
406—R. I. Hasell.
407—B. W. Andrews, Special Assistant Attorney-Gen., Washington, D. C.
408—H. L. Scaife, Lawyer, Clinton, S. C.
409—T. C. Dean, Broker, Spatranburg, S. C.
410—Palmer Brown, Director Chicago Crayon Company, Chicago, Ill.
411—J. G. Padgett, Lawyer, Walterboro, S. C.; Member Board Visitors The Citadel.
412—A. S. Manning, Bank President, Columbia, S. C.
413—J. J. Moorer.
414—A. S. Salley, Sec. and Lib. S. C. Historical Commission, Columbia, S. C.
415—David Hugenin, President Equitable Fire Insurance Company, Charleston, S. C.

Class of 1893

416—D. J. Lucas.
417—J. W. Cantey, Farmer, Boykin, S. C.
418—F. S. Wilcox, Electrical Engineer.
419—G. Shanklin, Assistant Professor Mathematics, Clemson College, S. C.
420—J. P. Thomas, Treasurer Cameron & Barkley Co.; Member Board of Visitors The Citadel, Charleston, S. C.
421—R. M. Perrin, Capt., A. G. Dept., N. A.
422—W. A. Stribling, Superintendent Cotton Mill, Union, S. C.
423—E. B. Fishburne, Headmaster Tennessee M. I., Sweetwater, Tenn.
424—B. D. Wilson, Lieutenant U. S. V.; Teacher, Sumter, S. C.

Class of 1893   Remarks

425—W. B. Gourdin.
426—G. H. Atkinson, President Albermarle N. & I. College, Albermarle, N. C.
427—J. H. Earle.
428—G. Bowen, Morris-Bowen Hardware Company, Birmingham, Ala.
429—W. E. Woodward, Banker, 200 W. 79th St., New York, N. Y.
430—G. L. Dial, Fire Insurance, Columbia, S. C.
431—J. R. Verdier, Lawyer, Utah.

Class of 1894

432—O. F. Hunter, Clerk Navy Department, Washington, D. C.
433—G. M. Stackhouse, Paymaster, Lt.-Com., U. S. N.
434—F. W. Gregg, Minister, Rock Hill, S. C.
435—T. E. L. Lipsey, Civil Engineer, Lincolnton, N. C.
436—W. P. Witsell, Minister, P. E. Church, Waco, Texas.
437—J. G. Johnston, Physician, Chester, S. C.
438—J. T. West, Bookkeeper, Cotton Oil Mill, Belton, S. C.
439—J. E. Peurifoy, Circuit Judge, Walterboro, S. C.
440—R. E. Babb, Lawyer, Laurens, S. C.
441—W. P. Odom, Merchant, Chesterfield, S. C.
442—C. C. Fishburne, Bookkeeper, Columbia, S. C.
443—W. S. Lee, Civil and Electrical Engineer, Charlotte, N. C.
444—R. H. McMaster, Colonel, U. S. A.
445—P. E. Hutto, Merchant, Swansea, S. C.
446—S. P. Anderson, Anderson Lumber Company, Charleston, S. C.
447—St. C. B. Gwynn.
448—E. H. Jeffords, Bookkeeper, Ice Del. Company, Charleston, S. C.
449—E. L. Ready, Farmer, Johnston, S. C.
450—T. C. Stevenson, Civil Engineer, Charleston, S. C.
451—J. W. Rouse, Teacher, Richland, Ga.
452—H. Horlbeck.
453—H. E. DePass, Lawyer, Spartanburg, S. C.
454—F. E. Hinnant, Cashier Bank, Sumter, S. C.
455—L. L. Gregory, Physician, Alcolu, S. C.
456—D. Kearney, Stenographer, Charleston, S. C.
457—S. J. DuPre, Cotton Mill Office, Glendale, S. C.
458—W. W. Clement, Superintendent Phosphate Company, Charleston, S. C.
459—W. St. J. Jervey, Maj. U. S. A.
460—A. E. Legare, Major, U. S. N. G.
461—B. R. Hiers, Lawyer, Hampton, S. C.
462—I. J. Burris, Physician, Anderson, S. C.
463—P. S. Norris, Aiken, S. C.
464—A. C. Baskin, Teacher, Bishopville, S. C.
465—G. M. Stuckey, Bank Official, Bishopville, S. C.
466—P. J. Peterkin, Farmer, Fort Motte, S. C.
467—J. A. Moroso, Literary Work, New Jersey.
468—J. E. Keith, Commercial Traveler, Cincinnati Shoe House.
469—W. G. Fike.
470—W. S. Langford, Wichita Falls, Texas; Captain U. S. Vol, 1898.

Class of 1894 — Remarks

471—J. D. Cozby, Captain, U. S. R.
472—T. C. Stone, Major Med. Corps.
473—E. C. Logan.
474—E. L. McIntosh.
475—E. A. McClellan, Physician, McClellanville, S. C.
476—W. K. Jackson.
477—E. L. Parker, Ph.D., Johns Hopkins; Professor Medical College of South Carolina, Charleston, S. C.
478—J. P. Smith, Lieutenant, U. S. N. R.
479—E. R. Thompkins, Col. U. S. A.
480—W. T. Green, Lawyer, Columbia, S. C.
481—R. L. Hughes, Teacher, Hampton County.

Class of 1895

482—S. W. Reeves, Professor Mathematics, Univ. of Okla., Norman, Okla.
483—H. C. Schwecke, Electrical Engineer, Pittsfield, Mass.
484—A. Levy, Lt.-Col., F. A., U. S. N. G.
485—P. T. Hayne, Col. U. S. A.
486—J. B. Allison, Colonel, U. S. A.
487—S. H. Booth, Minister, S. C. Conference, M. E. Church, South.
488—C. I. Green, Physician.
489—C. B. Smith, Col. C. A. C.
490—C. R. Harvin, Lumber Business, Manning, S. C.
491—J. B. Livingston, Railroad Office, Wilmington, N. C.
492—J. J. F. Barnes.
493—J. E. Minter, Bank Official, Laurens, S. C.
494—R. E. Boggs, Contractor, Spartanburg, S. C.
495—C. Martin, Wholesale Grocer, Wilmington, N. C.
496—P. Grausman, Physician, 120 W. 53rd Street, New York City.
497—C. T. Dowling, Merchant, Hix, S. C.
498—E. R. Wallace, Union, S. C.
499—C. Matheson, Lawyer, Gainesville, Fla.
500—H. A. Douglas, Asheville, N. C.
501—S. D. Jervey.
502—C. D. Rollins, Physician, Baltimore, Md.

Class of 1896

503—S. P. J. Garris, Cotton Oil Mill, Denmark, S. C.
504—B. G. Murphy, S. C. Conference, M. E. Church, South.
505—T. W. Carmichael, Physician, Bennettsville, S. C.
506—A. H. Marchant, Captain, U, S, R.
507—F. K. Holman, Physician, Sumter, S. C.
508—S. W. Carwile, Teacher, Ridge Spring, S. C.
509—E. J. Rogers, Suprintendent Vermont Sanatorium Pittsfield Vt.
510—S. M. Martin, Professor Mathematics, Clemson College, S. C.
511—J. P. Galvin, Physician, Charleston, S. C.
512—E. C. Wilcox.
513—P. A. McMaster, Lawyer, Columbia, S. C.

Class of 1896   Remarks

514—G. L. Dickson, Merchant, Lake City, S. C.
515—J. H. Taylor, Physician, Columbia, S. C.
516—J. S. Matthews, Physician, Denmark, S. C.
517—E. Croft, Col., U. S. A.
518—J. P. Guess, Farmer, Denmark, S. C.
519—H. G. Kaminer, Merchant, Gadsden, S. C.

Class of 1897

520—C. S. Bartless, Cotton, Shreveport, La.
521—R. D. Epps, Lawyer, Sumter, S. C.
522—E. C. Horton.
523—A. G. Holmes, Professor, Clemson College, S. C.
524—F. A. Coward. Capt., U. S. R., Med. Corps.
525—G. R. Fishburne, Broker, Charleston, S. C.
526—S. M. McLeod, Railway Mail Service, Rambert, S. C.
527—H. M. Langley, with Cr. Men's Pro. Assn., 615 Broadway, New York. N. Y.
528—J. D. Dial, Assistant Commissioner of Immigration, Columbia, S. C.
529—Roy Terrell, Railroad employ, Dallas, Texas.
530—R. J. Tillman, Col., U. S. A.
531—R. L. Stokes, Physician.
532—J. B. DuBose, Merchant, Marion, S. C.
533—Wm. Mazyck.
534—A. M. Deal, Lawyer, Columbia, S. C.
535—A. P. McElroy, Physician, Union, S. C.

Class of 1898

536—J. L. Fitts, Socialist Lecturer.
537—T. W. Bethea, General Agent New England Mutual Life Ins. Co., Charleston, S. C.
538—F. H. Derrick.
539—C. C. Derrick, Minister, M. E. Church, South, S. C. Conference.
540—J. J. Tuten, Farmer and Civil Engineer, Furman, S. C.

Class of 1899

541—S. O. Cantey, Minister, M. E. Church, S. C. Conference.
542—F. M. Ellerbe, Captain, C. A. C.
543—J. R. Crouch, Lawyer, Fort Motte, S. C.
544—A. Bramlett, Maj., C. A. C.
545—J. B. Salley, Lawyer, Aiken, S. C.
546—E. R. Price.
547—J. F. Townsend, Physician, Charleston, S. C.
548—W .F. Farmer, Manager Farmers' Oil Mill, Anderson, S. C.
549—S. C. Morris, Minister M. E. Church, South, S. C. Conference.

Class of 1900   Remarks

550—D. A. Bradham, Lawyer, Warren, Ark.
551—J. W. Linley, Real Estate, Anderson, S. C.
552—J. W. Moore, Commandant and Professor of History and Political Science, The Citadel, Charleston, S. C.; Maj. N. A.
553—W. E. Law, Bldg. Supplies, Charlotte, N. C.
554—W. W. Smoak, Proprietor and Editor, Press and Standard, Walterboro, S. C.
555—B. Calhoun, Assistant Superintendent Cotton Oil Mill, Clio, S. C.
556—C. W. DuRant.
557—A. J. Hydrick, Lawyer, Orangeburg, S. C.
558—J. R. Westmoreland, Assistant to President of Pacolet Manufacturing Company, Pacolet, S. C.
559—H. T. Rogers, Lawyer, Dyersburg, Tenn.
560—W. S. Clayton, Assistant Collector Internal Revenue, Wilmington, N. C.
561—J. H. Courtney, Farmer, Edgefield County.
562—W. H. Sligh, 216 Eighth Street, S. W., Washington, D. C.
563—J. P. Quarles, General Agent Equitable Life Ass. Co., Charlotte, N. C.
564—R. C. Bruce, Physician, Greenville, S. C.
565—L. M. Cochrane, Bookkeeper Bank, Anderson, S. C.
566—J. H. Haynesworth, County Superintendent of Education, Sumter, S. C.
567—S. C. Snelgrove, Paymaster, Lieut., U. S. N.
568—W. H. Evans, Teacher, Boys' High School, Atlanta, Ga.

Class of 1901

569—L. B. Steele, Lumber Business, Georgetown, S. C.
570—W. C. Hughs, Lawyer, Walhalla, S. C.
571—B. Kennedy.
572—W. F. Michau, Commercial Traveler, Charlotte, N. C.
573—T. M. Lyles, Lawyer, Spartanburg, S. C.
574—H. Hopkins.
575—E. B. Jackson, Bank Cashier, Wagener, S. C.
576—W. G. Martin, Commandant Mil. Acad., College Station, Texas.
577—H. D. Still, Merchant, Blackville, S. C.
578—W. C. O'Driscoll, Major, U. S. Med. Corps.
579—T. P. Lesesne, City Editor News and Courier, Charleston, S. C.
580—E. M. Allen, Physician, Florence, S. C.
581—E. C. Mann, Lawyer, St. Matthews, S. C.
582—C. S. McCall, Farmer, Bennettsville, S. C.
583—D. C. Pate, Captain, U. S. R.

Class of 1902

584—T. H. Russell, Headmaster, Staunton Military Academy, Staunton, Va.
585—C. C. Craft, Health Surgeon, S. S. Newport.
586—R. W. Wonson, Professor of History and Asst. Headmaster, Staunton Military Academy, Staunton, Va.
587—T. I. Weston, Civil Engineer, Columbia, S. C.
588—S. L. Bethea, Paymaster, U. S. N.; Lt.-Com.
589—J. W. Manuel, Lawyer, Hampton, S. C.
590—C. E. Daniel, Lawyer, Spartanburg, S. C.

Class of 1902   Remarks

591—E. E. Ballentine, Teacher, Long Ridge, S. C.
592—B. M. Thompson, Civil Engineer, A. C. L. Ry.
593—F. S. Muller, Teacher High School, Charleston, S. C.
594—W. C. White, Daily Report Examiner, Hartford Fire Ins Co., Atlanta, Ga.
595—T. E. Wilson, Civil Engineer, Darlington, S. C.
596—B. J. Robinson.
597—H. E. Raines, Manager Charleston Transfer Co., Charleston, S. C.
598—J. H. Thayer, Th.D., So. Bap. Theol. Sem., Minister, Lancaster, S. C.
599—T. C. Marshall, Civil Engineer, Chagrin Falls, Ohio.
600—L. A. McLeod.
601—E. E. Jenkins, Civil Engineer, New York City.
602—E. R. Tucker, Teacher, C. M. A., College Park, Ga.
603—A. H. Cross, Tampa, Fla.
604—L. N. Fishburne, Goldfield, Nev.
605—J. L. Gardner, Insurance, Fort Towson, Okla.
606—G. H. Miller, Civil Engineer.
607—J. R. Ashe, Physician, St. Luke's Hospital, New York, N. Y.
608—G. L. Rea, Physician, Snyder, Okla.
609—J. M. Beaty, Southern Express Company, Monroe, N. C.
610—A. T. Davis, Teacher, Mercersberg Academy.
611—S. F. Utsey.
612—W. E. Hutson, 1st Lieutenant Engineers, U. S. R.
613—D. K. Humphreys, Southern Express Company, Columbia, S. C.
614—J. Palmer, Civil Engineer, Sumter, S. C.
615—E. H. Smith, Shipping, Funch, Edy & Co., Maritime, N. Y.
616—T. J. Ashe, Electrical Engineer, 605 Victoria Ave., Westmount, Quebec, Canada.
617—W. B. Ravenel, Cotton, Charleston, S. C.
618—W. A. Klauber, Merchant, Bamberg, S. C.
619—E. N. Mittle, Cashier Bank, Bowman, S. C.

Class of 1903

620—D. G. Copeland, Civil Engineer, U. S. Naval Av.
621—R. F. McCracken, Asst Prof. Chem., Richmond Med Col.
622—I. A. Giles.
623—R. B. Cole, Captain, U. S. R.
624—C. E. Seybt.
625—W. G. Willard, Civil Engineer, Spartanburg, S. C.
626—A. E. Hutchison, Lawyer, Rock Hill, S. C.
627—W. A. Johnson, Merchant, North, S. C.
628—W. D. Watson.
629—W. B. Metts, with Planters' Fertilizer Company, Charleston, S. C
630—A. P. Barnes, Druggist, Walterboro, S. C.
631—J. H. McIlwinen, Farmer, Fayetteville, N. C.
632—J. M. Goodwin.
633—T. W. Hutson, Rice Planter, Yemassee, S. C.

Class of 1903   Remarks

634—L. Tiedeman, Wholesale Grocer, Charleston, S. C.
635—E. M. Tiller, Quartermaster, The Citadel.
636—K. R. McMaster, Merchant, Winnsboro, S. C.
637—H. A. Workman, Civil Engineer, Camden, N. J.

Class of 1904

638—G. L. Warren.
639—G. W. White, Civil Engineer, Charlotte, N. C.
640—J. T. Reese, Insurance, Columbia, S. C.
641—E. L. Culler, Farmer, Raymond, S. C.
642—W. E. Sawyer, Teacher, Miami, Fla.
643—C. M. Drummond, Lawyer, Woodruff, S. C.
644—N. P. Gettys, Camden, S. C.
645—L. J. Hammett, Physician, Greenville, S. C.
646—T. J. Lyon, Superintendent Schools, Edgefield, S. C.
647—J. F. O'Mara, Paymaster, U. S. N.
648—J. C. Hutchins, Liberty, S. C.
649—A. C. Padgett, Teacher.
650—A. L. Hodges, Captain, U. S. R.
651—E. L. Fishburne, Lawyer, Walterboro, S. C.
652—E. Iseman, Physician.
653—N. E. Rogers, Merchant, Florence, S. C.
654—W. L. Hemphill, Captain, U. S. R.
655—W. D. Acker, Principal Bolton College Agricultural High School, Brunswick, Tenn.
656—E. M. Kennedy, Merchant, Blackstock, S. C.

Class of 1905

657—R. F. Willingham, Cotton Factor, Macon, Ga.
658—L. W. Smith, Lieut., U. S. R.
659—E. C. Register, Lt.-Col., U. S. A. Med. Corps.
660—M. A. Hardnett, Electrical Engineer, Longview, Tex.
661—J. R. Cain, Pinopolis, S. C.
662—J. W. Martin, Captain Engineers, U. S. R.
663—H. A. Smith, Manager of Laundry, Florence, S. C.
664—R. E. Craig, Southern Oil Company, Columbia, S. C.
665—L. C. Still, U. S. Government Employ, Washington, D. C.
666—R. C. Dickson, Teacher, Westminster, S. C.
667—T. H. Moffatt, Captain, U. S. R.
668—R. B. Hartzog, Manager Sheridan's Teacher's Agency, Atlanta, Ga.
669—F. C. Easterby, 1st Lieut. Motor Truck Div., U. S. R.
670—W. M. Bostock, Civil Engineer, Aguas Calientes, Mexico.
671—W. M. Scott, Superintendent of Schools, Bishopville, S. C.
672—J. B. Doty, Merchant, Winnsboro, S. C.
673—Fitzhugh Lee, Druggist, Greenwood, S. C.
674—W. R. Richey, Captain, U. S. R.

Class of 1906   Remarks

675—F. B. Culley, Street Railway Company, Augusta, Ga.
676—J. J. McLure, Captain, C. A. C.
677—J. W. Simons, Jr., Major, U. S. A.
678—F. F. LaRoche, Draftsman, Atlanta, Ga.
679—F. G. Eason, Captain Engineers, U. S. R.
680—J. M. Moorer, Lawyer, Walterboro, S. C.
681—J. H. Johnson, Lawyer, Allendale, S. C.
682—G. M. Howe, Civil Engineer, Charleston, S. C.
683—R. D. Eadie, Teacher, Sparta, Ga.
684—C. C. Wyche, Major, U. S. R.
685—R. C. Moore.
686—J. R. Dickson, Captain, U. S. R.
687—R. W. Wingo.
688—F. H. McKinney, Teacher, Tigersville, S. C.
689—R. E. Gribben, Lieutenant, Chaplain, U. S. N. A.
690—P. J. Harrison.
691—C. F. Colvin, Business Manager Tulsa Democrat, Tulsa, Okla.
692—R. E. Corcoran, Paymaster, U. S. N.
693—W. W. Dick, Major, U. S. N. G.
694—J. L. M. Irby, Captain Engineers, U. S. N. G.
695—J. G. Lowry, Physician, New York.
696—H. G. Smith, with Cotton Manufacturing Company, Orangeburg, S. C.
697—J. O. Craig, Civil Engineer.
698—C. A. Roof.
699—F. A. Oakes, Private, Nineteenth Infantry, United States Army.
700—L. E. Langston, Civil Engineer, Dallas, Texas.
701—C. Waring.
702—W. P. Pollitzer, Sales Manager, Hartford Suspension Company, Jersey City, N. J.
703—W. A. Smith, Captain, U. S. R. Med. Corps.
704—J. E. McDonald, Lawyer, Winnsboro, S. C.
705—H. H. Stevens.
706—T. C. McGee, 1st Lieut. U. S. R.
707—F. G. Auld, Merchant, Eastover, S. C.
708—E. J. Blank, Lawyer, Charleston, S. C.
709—C. W. Muldrow, Captain, U. S. N. G.

Class of 1907

710—J. S. Bethea, Farmer, Latta, S. C.
711—W. W. Benson, Teacher, B. M. I., Greenwood, S. C.
712—W. D. Roper.
713—T. G. Russell, Commandant Staunton Military Academy, Staunton, Va.
714—B. H. Martin, Civil Engineer, Arkansas.
715—J. P. Clarke, Civil Engineer, Richmond, Va.
716—J. B. Hodges, Charleston Navy Yard.
717—R. C. Hunter, Prosperity, S. C.
718—W. T. Mikell, Traveling Salesman, Columbia, S. C.
719—W. J. Murray, Mercantile Business, Columbia, S. C.

Class of 1907

720—O. B. Hutson, in Business, Columbia, S. C.
721—Gordon Simmons, Electrical Engineer.
722—D. F. Bradham, Traveling Salesman, Jacksonville, Fla.
723—J. H. Hammond, Lawyer, Columbia, S. C.; Member Board of Visitors The Citadel.
724—J. C. Plowden, Asst. Cashier Bank, Manning, S. C.
725—P. S. Cromer, Civil Engineer, Atlanta, Ga.
726—T. D. Watkins, Lieut., N. A.
727—P. S. Connor, Real Estate, Atlanta, Ga.
728—J. G. Ehrlich, Mercantile Business, Columbia, S. C.

Class of 1908

729—R. H. Willis, Lt.-Col., U. S. A.
730—J. F. Nobrden, Principal Mitchell School, Charleston, S. C.
731—D. M. Myers, Principal Myers School, Savannah, Ga.
732—G. A. Townes, North Augusta School.
733—L. C. Bryan, 1st Lieut., U. S. N. A.
734—H. R. Wilkins, Insurance Business, Greenville, S. C.
735—A. P. McGee, Captain, U. S. R.
736—J. C. Pate, Teacher, Jefferson, S. C.
737—W. T. Briggs, Physician, North Augusta, S. C.
738—J. W. Campbell, Lt., N. A.
739—J. D. Charles, Bookkeeper, Greenville, S. C.
740—P. T. Palmer, 1st Lieut., U. S. R.
741—E. S. C. Baker, Lawyer, Conway, S. C.
742—H. R. Padgett, Lieut., N. A.
743—W. B. Porcher, Newspaper work.

Class of 1909

744—T. H. Rainsford.
745—W. D. Workman, Major, U. S. N. G.
746—C. L. Hair, Assistant Professor Mathematics, The Citadel, Charleston, S. C.
747—C. M. McMurray, Maj., U. S. A.
748—F. L. Link.
749—J. F. Muldrow, Captain, U. S. R.
750—R. M. Evans, Lieut., N. A.
751—L. K. Brown, in Bank, Florence, S. C.
752—M. B. Garris, United States Engineer, Jacksonville, Fla.
753—J. C. Busbee, Captain, U. S. R.
754—F. S. Smith, Teacher, Mayesville, S. C.
755—J. M. Lyles, in Bank, Winnsboro, S. C.
756—C. L. Harris, Instructor, Pennsylvania State College.
757—J. S. Nixon, Lieut., N. A.
758—H. A. Simms, Captain, U. S. R.
759—F. J. Watson, Civil Engineer, Bennettsville, S. C.
760—H. S. Haynesworth, Columbia, S. C.

Class of 1909   Remarks

761—A. Brunson, Lieut., U. S. N. G.
762—A. P. Rhett, Captain, U. S. R.
763—W. W. Barr.
764—C. K. McKie, Bank Teller, Augusta, Ga.
765—J. G. Osborne, Civil Engineer.
766—J. M. Sturgeon, Tobacco Business, Kentucky.
767—W. L. Reardon, Bookkeeper Bank, Graniteville, S. C.
768—S. L. Rigby, Capt., N. A.

Class of 1910

769—A. S. Harby, Lawyer, Sumter, S. C.
770—W. T. Lawton, Commandant, Donaldson Military School, Fayetteville, N. C.
771—L. R. Forney, Captain, U. S. N. G.
772—P. A. Clarke, Paymaster, Lt., U. S. N.
773—R. C. Williams, Major, U. S. A.
774—A. T. Cocoran, Y. M. C. A. Sec., U. S. N.
775—W. R. Conolly, Lt.-Col., U. S. A.
776—W. H. Langford, Teacher, Jonesville, S. C.
777—E. H. Huff, Lt., N. A.
778—W. C. Wylie, Equitable Life Insurance Society, Rock Hill, S. C.
779—B. C. Riddle, 1st Lieut., U. S. R.
780—J. W. Wallace, Teacher, Central, S. C.
781—W. W. McIver, Lieut., U. S. R.
782—F. P. Sessions, Major, U. S. R.
783—H. A. Woodward, Lawyer, Augusta, Ga.
784—S. L. Duckett, Civil Engineer, Chesterfield, S. C.
785—W. Q. Claytor, Merchant, Hopkins, S. C.
786—J. Rosenbaum, Greenwood, S. C.
787—J. R. Stewart, Capt, Engineers, U. S. R.
788—T. C. Parker, 1st Lieut., U. S. R.
789—G. C. Rogers, Principal Courtenay School, Charleston, S. C.
790—A. B. Gross, Atlantic Eng. Co., Savannah, Ga.
791—G. C. Blount, Lieut., U. S. R.
792—J. D. Parks, Lieut., N. A.
793—J. B. Grimball, 1st Lieut., U. S. R.
794—W. S. Lykes, Captain, U. S. R.
795—R. F. Bethea, Latta, S. C.
796—C. C. Wallace, Farmer, Kinards, S. C.
797—C. W. Reeves, New York, N. Y.
798—E. L. Skipper, 1st Lieut., Aviation Corps.
799—J. Laurens, 1st Sergt., U. S. N. G.
800—C. P. Cornwell, Lawyer, Tuscaloosa, Ala.
801—J. E. Cannon, U. S. Naval Res.
802—J. K. McCown, Lieut., N. A.
803—S. S. Tison, Lawyer, Barnettsville, S. C.
804—D. W. Gaston, Lawyer, Aiken, S. C.

Class of 1910   Remarks

805—E. D. Smith, Lieut., N. A.
806—**D. F. Fishburne.**
807—D. F. Moorer, St. George, S. C.
808—W. B. Stackhouse, Farmer, Latta, S. C.
809—E. C. Harvey, Farmer, Holly Hill, S. C.
810—C. M. Pilgrim, Woodruff, S. C.
811—W. M. Smith, Civil Engineer, 24 James Street, New York, N. Y.

Class of 1911

812—J. A. Lester, Major, U. S. A.
813—E. F. Witsell, Major, U. S. A.
814—S. A. Porter, Lieut., N. A.
815—G. W. Green, Private, N. A.
816—C. H. Fowler, Captain, N. A.
817—J. F. Risher, Teacher, Smoaks, S. C.
818—T. S. Sinkler, Jr., Captain, U. S. A.
819—C. A. Isaacs, Sumter Electric Works, Sumter, S. C.
820—H. G. Acker, Asst. Commandant, Staunton Military Academy, Staunton, Va.
821—J. K. Shannon, Gas and Water Company, Atlanta, Ga.
822—W. R. Buie, Civil Engineer, Georgetown, S. C.
823—W. R. Marvin, Farmer, White Hall, S. C.
824—B. T. Cripps, Major, U. S. M. C.
825—T. Street, Jr., in business, Savannah, Ga.
826—S. H. Clark, Real Estate Business, Savannah, Ga.
827—C. F. Yates, Civil Engineer, Texas.
828—F. A. Hazard, Architect, Wilson, N. C.
829—G. D. Murphey, Lieut.-Col., U. S. A.
830—R. E. Davis, 1st Lieut., U. S. R.
831—H. F. Porcher, Captain, U. S. N. G.
832—H. K. Pickett, Major, U. S. M. C.
833—G. C. McCelvey, Captain, U. S. R.
834—B. R. Legge, Major, U. S. A.
835—C. T. Smith, Jr., 1st Lieut., U. S. N. G.
836—B. A. Sullivan, Jr., Lieut., U. S. N. G.
837—H. O. Strohecker, Jr., Principal Bennett School, Charleston, S. C.
838—S. G. Thompson, Jr., in Bank, Abbeville, S. C.
839—J. C. Pickens, Civil Engineer, Charleston, W. Va.
840—J. E. Ellerbe, Jr., Civil Engineer, Winston-Salem, N. C.
841—B. D. Refo, Jr., Teacher, Lanes, S. C.
842—C. Johnson, Civil Engineer, Frankfort, Ind.

Class of 1912

843—S. S. Pitcher, Post Adjutant, Staunton Military Academy, Staunton, Va.
844—L. Simons, Lieut., U. S. A.
845—**A. C. Hiers.**
846—C. S. Brown, Hospital Corps, N. A.
847—**A. E. Merrimon.**
848—J. D. E. Meyer, Major, U. S. N. G.

Class of 1912   Remarks

849—E. B. Patrick, Lieut., U. S. R.
850—J. H. Bouknight, Lieut., N. A.
851—R. F. Walsh, Maj., U. S. A.
852—C. M. Lindsay, Major, U. S. R.
853—J. H. Thompson, 1st Lieut., C. A. C.
854—A. F. Littlejohn, 1st Lieut., U. S. N. G.
855—J. S. Sanders, Tobacco Business, Kentucky.
856—I. Riff, Georgetown, S. C.
857—J. C. Fair, 1st Lieut., U. S. R.
858—G. H. McLean, Lieut., N. A.
859—R. O. Free, Barnwell, S. C.
860—J. A. Doyle, Real Estate Business, Georgetown, S. C.
861—C. O. Kirsch, Lieut., N. A.
862—C. Rigby, Captain, Med. Corps, N. A.
863—S. E. Lyles, Insurance, Winnsboro, S. C.
864—O. G. Wood, Lieut., N. A.
865—C. Anderson, Jr., Engineer, Splitdorf Elec. Co., Sumter, S. C.
866—T. P. Duckett, 1st Lieut., U. S. N. G.
867—J. W. Shuler, Paymaster, U. S. N.
868—B. B. Bouknight, Farmer, Johnston, S. C.
869—J. C. Perrin, U. S. Interstate Commerce Commission, Lieut. Engineers, N. A.
870—W. H. Varn, Standard Oil Co., Hongkong, China.
871—F. Y. Legare, Farmer, Mullet Hall, S. C.
872—P. Robinson, Lieut., U. S. A.
873—J. M. Roper, Census Bureau, Washington, D. C.
874—M C. Stuckey, in Business, Florence, S .C.
875—J. P. Temple, Auditor Savannah Hotel, Savannah, Ga.
876—J. F. Oglesby, Draftsman.

Class of 1913

877—J. F. Hutchinson, Chemist.
878—H. E. Losse, 1st Lieut., U. S. R.
879—J. P. Woodson, 1st Lieut., Engineers, U. S. R.
880—J. M. Arthur, Major, U. S. M. C.
881—S. C. Chandler, Teacher and Y. M. C. A. Secretary, Staunton Military Academy, Staunton, Va.
882—R. N. Whaley, Assistant Secretary Committee Prevention Tuberculosis, 1701 Chestnut Street, Philadelphia, Pa.
883—D. S. DuBose, Lieut., N. A.
884—E. C. Hesse, Druggist, Charleston, S. C.
885—J. R. Martin, Captain, U. S. M. C., Killed in Santo Domingo, 1917.
886—C. P. Gilchrist, Major, U. S. M. C.
887—J. T. Yarborough, Captain, U. S. M. C.
888—J. R. Harris, Sergt., N. A.
889—L. A. Mims, Florence, S. C.
890—H. E. Sheldon, Lieut., N. A.

Class of 1913   Remarks

891—W. D. Boykin, Captain, U. S. R.
892—I. H. Kohn, Lieut., N. A.
893—J. W. Weeks, Captain, U. S. A.
894—A. S. LeGette, Captain, U. S. A.
895—M. W. Hester.
896—A. Smith, Captain, U. S. A.
897—H. C. Shirley, Honea Path, S. C.,
898—J. D. McDill, Stenographer, Columbia, S. C.
899—E. W. Marvin, 1st Lieut., U. S. A.
900—S. H. Smith, Private, Medical Corps.
901—B. D. Altman, Lieut., N. A.
902—W. H. Lawton, Ranchman, Montana.
903—F. W. Yates, Jr., Lieut., U. S. R.
904—J. C. Stanton, Captain, U. S. R.
905—A. M. Parrott, 1st Lieut., U. S. R.
906—D. F. Clark.
907—C. N. Muldrow, Captain, U. S. M. C.
908—L. W. Wilson, Captain, U. S. A.
909—P. L. Lybrand, in Business, Swansea, S. C.
910—D. F. Moore, Jr., in Business, Brunson, S. C.
911—J. C. Hutson, Captain, C. A. C.
912—T. K. Gibson, McColl, S. C.
913—L. W. Davis, Private, N. A.
914—E. P. Meadors.

Class of 1914

915—C. F. Myers, Jr., Asst. Prof. Mathematics, The Citadel.
916—H. T. Bridgman, Theological Student.
917—G. A. King, Captain, U. S. A.
918—J. Cart, Jr., 1st Lieut., U. S. R.
919—W. G. Thompson, in Business, 22 East Forty-fifth Street, New York, N. Y.
920—F. E. Harrison, Jr., Lieut., U. S. A.
921—W. E. Cuttino, in Business, Sumter, S. C.
922—E. W. Dabbs, Jr., 1st Lieut., U. S. R.
923—O. L. Long, Commandant, Sumter High School, Sumter, S. C.
924—I. Ussery, Insurance Business, Columbia, S. C.
925—A. B. Boykin, Boykin, S. C.
926—W. Prior, Paymaster, Lieut., U. S. N.
927—V. H. Wheeler, Paymaster, U. S. N.
928—T. F. McGarey, Captain Engineers, N. A.
929—T. H. Frost, Lieut., U. S. A.
930—P. J. Zeigler, Jr., Lieut., N. A.
931—N. Minus, Captain, U. S. A.
932—N. J. Smith, Principal High School, Rowesville, S. C.
933—J. W. Anderson, 1st Lieut., U. S. R.
934—S. A. Woods, Jr., Major, U. S. M. C.
935—L. W. Whaley, Major, U. S. M. C.

Class of 1914   Remarks

936—F. Y. Moore, Lieut., N. A.
937—K. I. Buse, Major, U. S. M. C.
938—S. F. Miller, with Southern Teachers' Agency, Columbia, S. C.
939—L. W. Boykin, Jr., Captain, U. S. R.
940—H. H.Gregory, Farmer, Fair Forest, S. C.
941—A. W. Lynch, Saluda, S. C.
942—J. F. Jeffords, Major, U. S. M. C.
943—S. L. Eason, Surveyor, American Shipping Bureau, New York.
944—H. B. Seyle, Captain, C. A. C.
945—G. E. Doyle, Lieut., U. S. R.
946—E. A. Sullivan, U. S. Engineers.
947—S. R. Moore, Teacher, Tirzah, S. C.
948—J. H. David, Jr., 1st Lieut., U. S. R.; Killed in France, 1918.
949—W. H. Flint, Draftsman, U. S. Navy Yard.
950—A. P. Bruner, Capt., U. S. A.
951—T. E. Hipp, Paymaster, U. S. N., Lieut.
925—A. H. Macaulay, Lieut., Air Service.
953—W. T. Barron, Private, N. A.

Class of 1915

954—T. B. Jackson, with Equitable Life Assurance Society, Rock Hill, S. C.
955—B. F. Gaines, Captain, U. S. N. G.
956—H. Tindal, Lieut., N. A.
957—W. C. Moore, Captain, U. S. A.
958—T. P. Cheatham, Major, U. S. M. C.
959—R. D. Porter, Captain, U. S. A.
960—G. A. Chalker, 1st Lieut., U. S. R.
961—J. H. Holmes, Jr., Captain, U. S. A.; killed in France in 1918.
962—H. J. Bailey, Lieut., Engineers, U. S. N. G.
963—C. S. Lawrence, Flying Cadet, A. S. A.
964—R. C. Hilton, Captain, U. S. A.
965—P. K. Shuler, Lieut., N. A.
966—R. D. Schroder, Teacher, Rockville, S. C.
967—H. O. Speed, 1st Lieut., F. A., N. A.
968—T. B. Baldwin, Lieut., N. A.
969—W. A. Moore, Jr., Lieut., C. A. C.
970—R. D. Hardy, Dupont Powder Company.
971—A. E. Nimitz, Draftsman, U. S. Navy Yard, Charleston, S. C.
972—I. B. Armfield, Teacher, Scranton, S. C.
973—R. W. Hudgens, Captain, U. S. N. G.
974—W. C. Wallace, Paymaster, U. S. N.
975—A. W. Folger, Captain, U. S. R.
976—R. L. Meares, Lieut., U. S. N. G.
977—G. W. Wilkes, U. S. N.
978—T. O. Cannon, Lieut., U. S. Aviation Corps.
979—H. Hutchison, Captain, U. S. N. G.
980—G. W. Clement, Paymaster, U. S. N.

Class of 1915

981—K. D. Ransom, Capt., U. S. M. C.
982—T. L. Alexander, Captain, U. S. A.
983—T. W. Martin, Lieut., N. A.
984—B. B. Kinloch, Captain, U. S. A.
985—J. W. Marshall, Sergt., U. S. Eng.
986—J. H. Morris, Lieut., N. A.
987—E. A. Terrell, Lieut., Chem. Warfare Service, N. A.
988—R. F. Boyd, Lieut., U. S. M. C.
989—R. L. Seay, Lieut., Signal Corps, N. A.
990—C. G. Hammond, Captain, U. S. A.
991—B. A. Grimball, Ensign, U. S. N. R.
992—S. L. Reid, Development Agent, S. A. L. Ry.
993—R. H. Tarrant, 1st Lieut., U. S. R.
994—E. J. Faylick, Hopkins, S. C.
995—R. J. Kirk, Captain, U. S. R.
996—J. W. Cooley, Lieut., U. S. R.
997—D. H. Owen, Major, U. S. M. C.

Class of 1916

998—J. A. Mood, Jr., Captain, U. S. A.; killed in France, 1918.
999—R. C. Brunson, Captain, U. S. A.
1000—G. A. Patrick, Captain, U. S. A.
1001—W. R. Cothran, Jr., Lieut., U. S. A.
1002—J. H. Lafitte, Captain, U. S. A.
1003—W. C. James, Major, U. S. M. C.
1004—T. D. Paulling, Lieut., N. A.
1005—W. M. Spann, Captain, U. S. A.
1006—J. Anderson, Lieut., U. S. N. G.
1007—F. W. Sheppard, Captain, U. S. A.
1008—W. C. Byrd, Major, U. S. M. C.
1009—W. M. Bouknight, Lieut., U. S. A.
1010—A. A. Cook, Captain, U. S. A.
1011—G. B. Reynolds, Major, U. S. M. C.
1012—D. H. Laird, Private, N. A.
1013—P. C. Pearson, Lieut., U. S. R.
1014—C. R. Perkins, Captain, U. S. R.
1015—F. J. Simons, Captain, U. S. A.
1016—H. C. Cooper, Major, U. S. M. C.
1017—J. K. Bolton, Lieut., U. S. M. C.; Killed in Santo Domingo, 1917.
1018—E. S. Blake, Captain, U. S. N. G.
1020—E. M. Claytor, with Bethlehem Steel Company, Newcastle, Del.
1021—A. R. Temple, Lieut., N. A.
1022—G. H. Yarbrough, Captain, U. S. M. C.; Killed in France, 1918.
1023—J. M. Gilbert, Lieut., N. A.
1024—T. P. Cothran, Jr., 1st Lieut., U. S. R.
1025—C. F. Kilgus, Bamberg, S. C.
1026—J. A. Gilbert, in Business, Wilmington, S. C.

Class of 1916   Remarks

1027—C. W. Chalker, Captain, U. S. A.
1028—J. T. Moore, Major, U. S. M. C.

Class of 1917

1029—J. G. McRea, Q. M. Dept., N. A.
1030—R. G. Howard, Lieut., U. S. M. C.
1031—J. F. Moriarty, Captain, U. S. M. C.
1032—W. Q. Jeffords, Captain, C. A. C., Schofield Barracks.
1033—E. W. King, Captain, C. A. C., Regular Army.
1034—W. W. Muckenfuss, Sergeant Engineers, N. A.
1035—J. L. Weeks, 1st Lieut., Infantry, Regular Army.
1036—E. B. Hope, Captain, U. S. M. C.
1037—J. W. Lea, Lieut., U. S. M. C.
1038—K. Green, 1st Lieut., Infantry. Regular Army.
1039—T. B. Fowler, Capt., Infantry, Regular Army.
1040—H. C. Switzer, 1st Lieut., Infantry, Regular Army.
1041—H. H. Jeter, Lieut., U. S. M. C.
1042—G. I. Chumbley, Lieut., U. S. M. C.
1043—W. K. Dickeson, 1st Lieut., Infantry, Regular Army.
1044—E. H. Poulnot, Lieut., U. S. M. C.
1045—H. C. Savage, Captain, U. S. M. C.
1046—D. A. Holladay, Capt., U. S. M. C.
1047—J. A. Clarkson, 1st Lieut., Infantry, N. A.
1048—S. Y. Dinkins, 1st Lieut., Infantry, Regular Army.
1049—J. A. Nichols, 1st Lieut., Cavalry, Regular Army.
1050—L. G. Merritt, Captain, U. S. M. C.
1051—W. A. Moore, Lieut., N. A.
1052—F. S. Poulnot, Lieut., N. A.
1053—O. C. Moore, Capt., C. A. C.
1054—E. M. Galphin, 1st Lieut., Infantry, National Army.
1055—G. H. Whisenhunt, Capt., U. S. M. C.
1056—E. P. Norwood, Captain, U. S. M. C.
1057—W. G. Wallace, Lieut., Field Artillery, Regular Army.
1058—J. C. Cogswell, Captain, U. S. M. C.
1059—J. P. Mahaffey, Private, N. A.
1060—A. T. Elmore, 1st Lieut., U. S. M. C.; killed in France, 1918.
1061—S. C. Strohecker, 1st Lieut., N. A.

Class 1918

1062—G. G. Cromer, Lieut., Inf., N. A.
1063—N. J. Cromer, Lieut., F. A., N. A.
1064—F. R. Rogers, Lieut., F. A., N. A.
1065—B. R. Stroup, Aviation Corps.
1066—H. F. Adickes, Lieut., U. S. M. C.
1067—W. R. Mood, Flying Cadet.
1068—R. Y. Turner, Lieut., F. A., N. A.
1069—W. N. Levin, Lieut., Inf., N. A.

Class of 1918   Remarks

1070—H. L. Cunningham, Sergeant, U. S. M. C.
1071—J. L. Bolt, Lieut., Inf., N. A.
1072—W. L. McKittrick, Lieut., U. S. M. C.
1073—F. E. Zemp, Private, Inf., N. A.
1074—W. P. Bowers, Lieut., Inf., N. A.
1075—T. C. Sparks, Lieut., Inf., N. A.
1076—L. C. Waring, Lieut., Inf., N. A.
1077—F. L. Gaffney, Ensign, Paymaster, U. S. N.
1078—J. E. White, Lieut., F. A., N. A.
1079—J. L. Dicks, Lieut., Inf., N. A.
1080—H. W. Carter, Sergt., U. S. M. C.
1081—W. J. Wallis, Lieut., Inf., N. A.
1082—F. N. Thurston, Lieut., Inf., N. A.
1083—H. E. Platt, Private, U. S. M. C.
1084—J. B. Gambrell, Lieut., U. S. M. C.
1085—B. C. Boland, Lieut., Inf., N. A.
1086—A. Middleton, Lieut., Inf., N. A.
1087—H. W. Tarkington, Lieut., Inf., U. S. A.
1088—K. L. Simons, Lieut., U. S. M. C.
1089—K. F. Snearer, Lieut., F. A., U. S. A.

HONOR ROLL

GRADUATES OF THE CITADEL IN THE MILITARY AND NAVAL SERVICES.

ARRANGED ACCORDING TO CLASSES

Symbols: *, indicates service in American Expeditionary Force; R. Regular Army; G, National Guard; N, National Army; V, Navy; M, Marine Corps.

1886	Edward Anderson	N	Major	Staff
	*L. S. Carson	R	Lt. Col.	Cav.
	*W. F. Robertson	G	Major	Inf.
1889	*W. W. Lewis	G	Lt. Col.	Inf.
	M. L. Smith	N	Major	J. Adv.
1890	W. H. Simons	R	Colonel Inf.	Died 1918
1891	E. M. Blythe	N	Major	Inf.
	J. D. Frost	N	Major	A. Gen.
	*P. K. McCully	G	Colonel	Inf.
	*A. M. Brailsford	G	Major	Med. C.
1893	R. M. Perrin	N	Captain	A. Gen.
1894	G. M. Stackhouse	V	Lt. Com.	Pay.
	*R. H. McMaster	R	Colonel	F. A.
	W. S. Jervey	R	Major	Inf.
	A. E. Legare	N	Major	Inf.
	*T. C. Stone	N	Major	Med. C.
	J. P. Smith	V	Lieutenant	Naval Res.
	E. R. Tompkins	R	Colonel	Cav.
1895	A. Levy	N	Lt. Colonel	F. A.
	*P. T. Hayne	R	Colonel	Cav.
	J. B. Allison	R	Colonel	Inf.
	*C. B. Smith	R	Colonel	C. A. C.
1896	*A. H. Marchant	N	Captain	Inf.
	*E. Croft	R	Colonel	Inf.
	*F. A. Coward	N	Captain	Med. C.
	B. J. Tillman	R	Colonel	Inf.
1899	F. M. Ellerbe	G	Captain	C. A. C.
	*A. Bramlet	N	Major	C. A. C.
1900	J. W. Moore	N	Major	Inf.
	S. C. Snelgrove	V	Lieutenant	Pay.
1901	*W. C. O'Driscoll	N	Major	Med. C.
	D. C. Pate	N	Captain	Inf.
1092	S. L. Bethea	V	Lieut. Com.	Pay
1903	D. G. Copeland	V	C. E.	Naval Av.
	*R. B. Cole	N	Captain	Inf.
1904	J. F. O'Mara	V	Com.	Pay.

107

Symbols: *, indicates service in American Expeditionary Force; R. Regular Army; G. National Guard; N, National Army; V, Navy; M, Marine Corps.

	A. L. Hodges	N	Captain	Ord.
	W. L. Hemphill	N	Captain	Inf.
1905	L. W. Smith	N	Lieutenant	Inf.
	E. C. Register	R	Lt. Colonel	Med. C.
	J. W. Martin	N	Captain	Eng.
	*T. H. Moffatt	N	Captain	Inf.
	F. C. Easterby	N	Lieutenant	Inf.
	*W. R. Richey	N	Captain	Inf., Wounded
1906	*J. J. McLure	N	Captain	C. A. C.
	*J. W. Simons	R	Major	Inf.
	F. G. Eason	N	Captain	Eng.
	*C. C. Wyche	N	Major	Inf.
	J. R. Dickson	N	Captain	Inf.
	*R. E. Gribben	G	Lieutenant	Chaplain
	R. E. Corcoran	V	Com.	Pay.
	W. W. Dick	G	Major	Inf.
	J. L. M. Irby	N	Captain	Eng.
	F. J. Oakes	R	Private	Inf.
	*W. A. Smith	N	Captain	Med. C.
	T. C. McGee	N	Lieutenant	Inf.
	C. W. Muldrow	G	Captain	Inf.
1907	T. D. Watkins	N	Lieutenant	Inf.
1908	*R. H. Willis	R	Lt.-Col., killed in France, 1918.	Avia.
	L. C. Bryan	N	Lieutenant	Inf.
	*A. P. McGee	N	Captain	Inf.
	J. W. Campbell	N	Lieutenant	Inf.
	P. T. Palmer	N	Lieutenant	Inf.
	H. R. Padgett	N	Lieutenant	Inf.
1909	*W. D. Workman	G	Major	Inf.
	*C. M. McMurray	R	Major	Inf.
	F. L. Link		Captain	Phil. Constab.
	J. F. Muldrow		Captain	Inf.
	R. M. Evans	N	Lieutenant	Inf.
	*J. C. Busbee	N	Captain	Inf.
	J. S. Nixon	N	Lieutenant	Inf.
	*H. A. Simms	N	Captain	Inf.
	A. Brunson	G	Lieutenant	Inf.
	*A. P. Rhett	N	Captain	Inf.
	S. L. Rigby	N	Captain	Inf.
1910	L. R. Forney	G	Captain	Inf.
	P. A. Clarke	V	Lieutenant	Pay
	R. C. Williams	R	Major	Y. M. C. A.
	*A. T. Corcoran	V		Y. M. C. A.
	*W. R. Conoley	R	Lieut.-Col.	F. A.
	E. H. Huff	N	Lieutenant	Inf.
	*B. C. Riddle	N	Lieutenant,	Inf., Wounded
	W. W. McIver	N	Lieutenant	Inf.
	*F. P. Sessions	N	Major	F. A.

Symbols: *, indicates service in American Expeditionary Force; R. Regular Army; G, National Guard; N, National Army; V, Navy; M, Marine Corps.

	J. R. Stewart ....N	Captain .... Eng.
	T. C. Parker ....N	Lieutenant .... Inf.
	G. C. Blount ....N	Lieutenant .... Inf.
	J. D. Parks ....N	Lieutenant .... Inf.
	*J. B. Grimball ....N	Lieutenant ....F. A.
	W. S. Lykes ....N	Captain .... Inf.
	*E. L. Skipper ....N	Lieutenant ....Av. C.
	*J. Laurens ....G	1st Sergeant .... Cav.
	J. E. Cannon ....N	............Naval Reserve
	J. K. McCown ....N	Lieutenant .... Inf.
	E. D. Smith ....N	Lieutenant .... Inf.
1911	*J. A. Lester ....R	Major ....F. A.
	E. F. Witsell ....R	Major .... Inf.
	S. A. Porter ....N	Lieutenant .... Inf.
	G. W. Green ....N	Private .... Inf.
	C. H. Fowler ....N	Captain .... Inf.
	T. S. Sinkler ....R	Captain .... Inf.
	B. T. Cripps ....M	Major ....U. S. M. C.
	G. D. Murphey ....R	Lieut.-Colonel .... Inf.
	*R. E. Davis ....N	Lieutenant .... Inf.
	*H. F. Porcher ....G	Captain ....Cav.
	H. K. Pickett ....M	Major ....U. S. M. C.
	*G. C. McCelvey ....N	Captain, Inf., Wounded, D. S. C.
	*B. R. Legge ....R	Major, Croix de Guerre; Legion of Honor .... Inf.
	*C. T. Smith ....G	Lieutenant .... Inf.
	*B. A. Sullivan ....G	Lieutenant .... Inf.
1912	*L. Simons ....R	Lieutenant .... Inf.
	C. S. Brown ....N	.... Hos. C.
	*J. D. E. Meyer ....G	Major .... Inf.
	E. B. Patrick ....N	Lieutenant ....C. A. C.
	J. H. Bouknight ....N	Lieutenant .... Inf.
	*R. F. Walsh ....R	Major .... Inf.
	C. M. Lindsay ....N	Major .... Inf.
	J. H. Thompson ....N	Lieutenant ....C. A. C.
	*A. F. Littlejohn ....G	Lieutenant ....Cav.
	J. C. Fair ....N	Lieutenant .... Inf.
	G. H. McLean ....N	Lieutenant .... Inf.
	C. O. Kirsch ....N	Lieutenant ....Med. C.
	C. Rigby ....N	Captain ....Med. C.
	O. G. Wood ....N	Lieutenant .... Inf.
	T. P. Duckett ....G	Lieutenant .... Inf.
	J. W. Shuler V	Ensign .... Pay.
	J. C. Perrin ....N	Lieutenant .... Eng.
	*P. B. Robinson ....R	Lieutenant .... Inf.
1913	*H. E. Losse ....N	Lieutenant .... Inf.
	*J. P. Woodson ....N	Lieutenant .... Eng.
	J. M. Arthur ....M	Major ....U. S. M. C.
	D. S. DuBose ....N	Lieutenant ....F. A.

Symbols: *, indicates service in American Expeditionary Force; R. Regular Army; G, National Guard; N, National Army; V, Navy; M, Marine Corps.

J. R. Martin	M	Captain, Died in Santo Domingo, 1917 .....U. S. M. C.
*C. P. Gilchrist	M	Major .....U. S. M. C.
J. T. Yarborough	M	Captain .....U. S. M. C.
J. R. Harris	N	Sergeant .....Inf.
H. E. Sheldon	N	Lieutenant .....Inf.
W. D. Boykin	N	Captain .....Inf.
I. H. Kohn	N	Lieutenant .....Inf.
*J. W. Weeks	R	Captain .....Cav.
*A. S. LeGette	R	Captain .....Inf.
A. Smith	R	Captain .....Inf.
E. W. Marvin	R	Lieutenant .....C. A. C.
S. H. Smith	N	Private .....Med. C.
B. D. Altman	N	Lieutenant .....Inf.
E. W. Yates	N	Private .....Inf.
J. C. Stanton	N	Captain .....Inf.
A. M. Parrott	N	Lieutenant .....Inf.
*C. N. Muldrow	M	Lieutenant .....U. S. M. C.
*L. W. Wilson	R	Captain .....Inf.
J. C. Hutson	R	Captain .....C. A. C.
L. W. Davis	N	Private .....Inf.
1914 G. A. King	R	Captain .....Cav.
J. Cart	N	Lieutenant .....Inf.
F. E. Harrison	R	Lieutenant .....C. A. C.
E. W. Dabbs	N	Lieutenant .....Inf.
W. Prior	V	Lieutenant .....Pay
V. H. Wheeler	V	Lieutenant .....Pay
*T. F. McGarey	G	Captain .....Eng.
*T. H. Frost	R	Lieutenant .....Inf.
J. J. Zeigler	N	Lieutenant .....Inf.
N. Minus	R	Captain .....Inf.
J. W. Anderson	N	Lieutenant .....Inf.
S. A. Woods	M	Captain .....U. S. M. C.
L. W. Whaley	M	Captain .....U. S. M. C.
F. Y. Moore	N	Lieutenant .....Inf.
K. I. Buse	M	Major .....U. S. M. C.
L. W. Boykin	N	Captain .....Inf.
J. F. Jeffords	M	Captain .....U. S. M. C.
H. B. Seyle	G	Captain .....C. A. C.
G. E. Doyle	N	Lieutenant .....Inf.
*E. A. Sullivan	G	Private .....Eng.
*J. H. David, Jr.	N	Lieut., killed in France, 1918. Inf.
A. P. Bruner	R	Captain .....C. A. C.
T. E. Hipp	V	Lieutenant .....Pay
A. H. Macaulay	N	Lieutenant .....A. S. A.
*W. T. Barron	N	Private .....Inf.
1915 B. F. Gaines	G	Captain .....Inf.
H. Tindal	N	Lieutenant .....Inf.

Symbols: *, indicates service in American Expeditionary Force; R. Regular Army; G, National Guard; N, National Army; V, Navy; M, Marine Corps.

W. C. Moore	R	Captain .... Inf.
T. P. Cheatham	M	Major ....U. S. M. C.
*R. D. Porter	R	Captain .... Inf.
G. A. Chalker	N	Lieutenant .... Inf.
*J. H. Holmes, Jr.	R	Captain, killed in France, 1918, decorated .... Inf.
*H. J. Bailey	N	Lieutenant .... Eng.
C. S. Lawrence	N	Flying Cadet ....A. S. A.
R. C. Hilton	R	Captain,...... Inf., Croix de Guerre
P. K. Shuler	N	Lieutenant .... Inf.
H. O. Speed	N	Lieutenant .... F. A.
T. B. Baldwin	N	Lieutenant .... Inf.
*W. A. Moore	R	Lieutenant ....C. A. C.
*R. W. Hudgens	G	Captain .... Inf., Wounded
W. C. Wallace	V	Lieutenant .... Pay.
A. W. Folger	N	Captain .... Inf.
R. L. Mearss	G	Lieutenant .... Inf.
G. W. Wilkes	V	Private .... Navy
T. O. Cannon	N	Lieutenant .... Av. C.
*H. Hutchinson	G	Captain .... Inf., Wounded
G. W. Clement	V	Lieutenant .... Pay.
K. D. Ransom	M	Captain ....U. S. M. C.
*T. L. Alexander	R	Captain .... Inf.
T. W. Martin	N	Lieutenant .... Inf.
*B. B. Kinloch	R	Captain .... Inf.
*J. W. Marshall	N	Sergeant .... Eng.
J. H. Morris	N	Lieutenant .... Inf.
E. A. Terrell	N	Lieutenant ....Chem. W. S.
R. F. Boyd	M	Lieutenant ....U. S. M. C.
R. L. Seay	N	Lieutenant .... Sig. C.
*C. G. Hammond	R	Captain .... Inf.
B. A. Grimball	V	Ensign .... Navy
*R. H. Tarrant	N	Lieutenant .... Inf., Wounded
R. J. Kirk	N	Captain .... Inf.
J. W. Cooley	N	Lieutenant .... Inf.
D. H. Owen	M	Major ....U. S. M. C.
1916 *J. A. Mood, Jr.	R	Captain, Inf., Decorated, killed in France, 1918.
R. C. Brunson	R	Captain .... Inf.
G. A. Patrick	R	Captain .... C. A. C.
W. R. Cothran	R	Lieutenant .... Inf.
J. H. Laffitte	R	Captain .... C. A. C.
W. C. James	M	Major ....U. S. M. C.
T. D. Pauling	N	Lieutenant .... Inf.
*W. M. Spann	R	Captain .... Inf.
*J. Anderson	G	Lieutenant ....Inf., Wounded
*F. W. Sheppard	R	Captain .... F. A.
W. C. Byrd	M	Major ....U. S. M. C.

111

Symbols: *, indicates service in American Expeditionary Force; R. Regular Army; G, National Guard; N, National Army; V, Navy; M, Marine Corps.

*W. M. Bouknight	R	Lieutenant ....Inf.
*A. A. Cook	R	Captain ....Inf.
G. B. Reynolds	M	Major ....U. S. M. C.
*D. H. Laird	N	Private ....Inf., Wounded
*P. C. Pearson	N	Lieutenant ....Inf., Cited for bravery
C. R. Perkins	R	Captain ....Inf.
F. J. Simons	R	Captain .... Cav.
H. C. Cooper	M	Major ....U. S. M. C.
J. K. Bolton	M	Lieutenant, U. S. M. C., killed in Santo Domingo, 1917.
*E. S. Blake	G	Captain ....Inf.
*O. A. Palmer	R	Captain .... Cav.
A. R. Temple	N	Lieutenant ....Inf.
G. H. Yarborough	M	Lieutenant, U. S. M. C., killed in France, 1918.
J. M. Gilbert	N	Lieutenant ....Inf.
T. P. Cothran	N	Lieutenant ....Inf.
*C. W. Chalker	R	Captain ....F. A.
J. T. Moore	M	Major ....U. S. M. C.
S. C. Strahecker	N	Lieutenant ....Inf.
1917 J. G. McRae	N	..........................Q. M. Dept.
R. G. Howard	M	Lieutenant ....U. S. M. C.
*J. F. Moriarty	M	Captain ....U. S. M. C.
W. Q. Jeffords	R	Captain ....C. A. C.
*E. W. King	R	Captain ....C. A. C.
*W. W. Muckenfuss	N	Sergeant ....Eng.
*J. L. Weeks	R	Lieutenant ....Inf., Wounded
*E. B. Hope	M	Captain, U. S. M. C., wounded; cited for bravery.
J. W. Lea	M	Lieutenant ....U. S. M. C.
*K. Green	R	Lieutenant ....Inf.
*T. B. Fowler	R	Captain ....Inf.
*H. C. Switzer	R	Lieutenant ....Inf.
H. H. Jeter	M	Lieutenant ....U. S. M. C.
G. L. Chumbley	M	Lieutenant ....U. S. M. C.
*W. K. Dickson	R	Lieutenant ....Inf.
E. H. Poulnot	M	Lieutenant ....U. S. M. C.
*H. C. Savage	M	Captain ....U. S. M. C.
*D. A. Holladay	M	Lieut., U. S. M. C., wounded twice.
*J. A. Clarkson	N	Lieutenant ....Inf.
S. Y. Dinkins	R	Lieutenant ....Inf.
*J. A. Nichols	R	Lieutenant .... Cav.
*L. G. Merritt	M	Captain ....U. S. M. C.
W. A. Moore	N	Lieutenant ....Inf.
F. S. Poulnot	N	Lieutenant ....Inf.
*O. C. Moore	R	Captain ....C. A. C.
*E. M. Galphin	N	Lieutenant ....Inf.
G. H. Whisenhunt	M	Captain ....U. S. M. C.

Symbols: *, indicates service in American Expeditionary Force; R. Regular Army; G, National Guard; N, National Army; V, Navy; M, Marine Corps.

*E. P. Norwood	M	Captain, U. S. M. C., Wounded, D.S.C.
W. G. Wallace	N	Lieutenant .... F. A.
*J. C. Cogswell	M	Captain, U. S. M.C., Severely wounded; D. S. Cross.
J. P. Mahaffey	N	Private .... Inf.
*A. T. Elmore	M	Lieutenant, U. S. M. C., killed in France.
1918 *G. G. Cromer	N	Lieutenant .... Inf.
N. J. Cromer	N	Lieutenant .... F. A.
F. R. Rogers	N	Lieutenant .... F. A.
B. R. Stroup	N	Pvt. 1st Cl., ... Av. C., A. S. A.
H. F. Adickes	M	Lieutenant .... U. S. M. C.
W. R. Mood	N	Pvt. 1st Cl. ... Flying Cadet, A. S. A.
R. Y. Turner	N	Lieutenant .... F. A.
W. N. Levin	N	Lieutenant .... Inf.
H. L. Cunningham	M	Sergeant .... U. S. M. C.
J. L. Bolt	N	Lieutenant .... Inf.
W. L. McKittrick	M	Lieutenant .... U. S. M. C.
F. E. Zemp		Private .... Inf.
W. P. Bowers	N	Lieutenant .... Inf.
T. C. Sparks	N	Lieutenant .... Inf.
L. C. Waring	N	Lieutenant .... Inf.
*F. L. Gaffney	V	Ensign .... Pay.
J. E. White	N	Lieutenant .... F. A.
J. L. Dicks	N	Lieutenant .... Inf.
*H. W. Carter	M	Sergeant .... U. S. M. C.
W. J. Wallis	N	Lieutenant .... Inf.
E. N. Thurston	N	Lieutenant .... Inf.
H. E. Platt	M	Private .... U. S. M. C.
J. B. Gambrell	M	Lieutenant .... U. S. M. C.
B. C. Boland	N	Lieutenant .... Inf.
A. Middleton	N	Lieutenant .... Inf.
H. W. Tarkington	R	Lieutenant .... Inf.
K. L. Simons	M	Lieutenant .... U. S. M. C.
*K. F. Snearer	N	Lieutenant .... F. A.
1919 W. C. Huggins	N	Lieutenant .... Inf.
T. C. Cannon	N	Lieutenant .... Inf.
J. J. Still	N	Lieutenant .... Inf.
J. H. Coleman	N	Lieutenant .... Inf.

EX-CADETS OF THE CITADEL IN THE MILITARY OR NAVAL SERVICE OF THE UNITED STATES

Symbols: *, indicates service in American Expeditionary Force; R. Regular Army; G, National Guard; N, National Army; V, Navy; M, Marine Corps.

Name	Class	
Adams, W. W.	1914	Aero Squad
Alexander, Moses	1920	2nd Lieut. Inf.
Alexander, Thos. L.	1909	Captain, National Army
Allan, G. H.	1920	2nd Lieut. Inf.
Allein, R. B.	1911	1st Lieut. Hq. Div.
Ancrum, A. S.	1907	Captain, Q. M. C.
Antley, E. B.	1914	Ensign, Naval Reserve.
Appleby, Jos.	1917	Navy. Died Oct. 1918.
Ardrey, W. B.	1921	2nd Lieut. Inf.
*Arthur, J. D.	1911	Major, U. S. A. Reg.
Ashe, P. H.	1918	U. S. N.
Atkinson, J. A.	1917	U. S. N.
Bailey, G. C.	1915	Lieut. 118 Inf.
Baker, B. R.	1919	Lieutenant Inf. N. A.
Bardin, T. J.	1918	Private National Army
Bell, F. D.	1918	Signal Corps
Bethea, Palmer	1912	Lieut. National Army
Bethea, W. T.	1919	Seregant Hos.
Boatright, C. B.	1916	Private National Army
Boineau, L. Calhoun	1916	2nd Lieut. U. S. A., Reg.
Bollin, J. H.	1913	Lieut. National Army.
Boyd, F. H.	1915	Hospital Corps, U. S. N.
Boykin, L. D.	1917	Lieut., Inf.
Brice, W. O.	1921	2nd Lieut. Inf.
Broadwater, A. L.	1908	Captain M. G. Co.
*Brown, G. L.	1914	2nd Lieut., Air Service
Brown, J. R.	1916	Corp. F. A.
*Brown, S. K.	1898	1st Lieut. Inf.
Bruce, J. G.	1919	Sergeant Inf.
Buck, H. H.	1919	Lieut., Inf., N. A.
Byers, E. D.	1918	Lieutenant Inf.
*Carrington, W. S.	1914	Lieutenant (S. G.) U. S. N.
Carroll, A. T.	1917	Seregant Hos.
Carroll, D. M.	1919	Aero Squad
Cave, L. A.	1913	Wagoner, Inf.
Cheatham, R. B.	1897	Captain, Inf.
*Childs, R. G.	1917	1st Lieut., Inf.
Childs, St. J. R.	1916	Lieutenant, U. S. M. C.
*Cochran, R. J.	1917	Lt. A.S.S. R.C., killed in France, 1918
Cogswell, W. H., Jr.	1913	Captain, Am. Tr.

Symbols: *, indicates service in American Expeditionary Force; R. Regular Army; G, National Guard; N, National Army; V, Navy; M, Marine Corps.

*Coleman, J. W.	1912	Lieutenant, Inf., Reg.
*Coleman, W. Osce	1915	Lieut., Inf., Reg., Croix de Guerre.
Collins, R. W.	1899	Colonel, C. A. C. Reg.
Cook, J. E.	1918	Aero Squad
Cooner, R. H.	1912	Sergeant-Major, Inf.
Cooper, T. B.	1919	Private, N. A.
Cope, G. W.	1912	Ensign, U. S. N.
*Cordes, C. E.	1918	Lieutenant, N. A.
Crouch, J. C.	1920	Private, Aero C.
Cunningham, T. H.	1898	Lieut.-Col., U. S. A. Reg.
Daniel, C. E.	1918	Lieutenant, N. A.
de Saussure, H. W.	1893	Capt. Med. Corps.
de Saussure, H. W., Jr.	1918	Lieutenant, Inf., N. A.
Dicks, R. V.	1920	O. T. S. Died Oct. 1918.
Donaldson, T. Q.	1916	2nd Lieut. U. S. A.
*Doolittle, T. B.	1917	Lieutenant, N. A.
DuBose, H. G.	1909	Lieutenant, N. A.
Dunkin, W. W.	1921	2nd Lieut. Inf.
Dunston, C. J.	1919	Avia. C.
Eason, J. D.	1919	Lieut. M. G. Co.
Edwards, A. C.	1921	U. S. N.
Ehrlich, F. N.	1905	Captain, Hq. Div.
Fieldet, T. F.	1918	Lieutenant, Air Service
Finley, W. G.	1915	Lieut. Inf., N. A.
Foy, G. N.	1918	Lieut. Inf., N. A.
Gamble, W. G.	1918	Sergeant Ord. Det., N. A.
Gibson, A. T.	1917	
Goldsmith, G. B.	1916	Lieutenant Inf., N. A.
Graham, J. C.	1918	U. S. S Arval
Gregorie, E. M.	1912	Lieutenant Inf., N. A.
*Gausp, A. P.	1920	Avia. Corp.
Hagood, Johnson	1917	Bugler, Inf., N. A.
Hahn, H. H.	1913	Lieutenant, Inf., N. A.
*Hankey, H. B.	1919	Avia. Corp.
Hardison, K. M.	1914	Captain, N. G.
Hardwicke, G. W.	1916	First Lieutenant, N. A.
Hare, L R.	1917	U. S. N.
Harllee, W. C.	1895	Major, U. S. M. C.
Hayne, T. B.	1920	Avia. Corp.
Haynsworth, J. H.	1886	Ensign, U. S. N.
Hildebrand, H. Z.	1914	Sergeant, Inf., N. A.
Hill, A. B.	1917	Lieutenant, Inf., N. A.
*Hoke, G. M.	1916	Lieutenant, Inf., N. A.
Hunter, M. R.	1914	Signal Corps.
Jeffords, R. L.	1920	Lieutenant, Inf., N. A.
Jeffries, J. F.	1920	Signal Corps, U. S. N.
*Johnson, H. L.	1921	Private, U. S. Eng.
*Johnson, J. B., Jr.	1916	Sergeant, Am. Tr., N. A.

Symbols: *, indicates service in American Expeditionary Force; R. Regular Army; G, National Guard; N, National Army; V, Navy; M, Marine Corps.

Johnson, K. E.	1919	Lieut. M. G. Co., N. A.
Kelly, J. D.	1910	Major, Cav., Reg.
Klein, A. L.	1911	Lieutenant, Inf., N. A.
Lake, P.	1917	Tr. Camp, Inf.
*Lake, T. D.	1914	Lt., Inf., N. A., killed in France 1918.
Lane, R. L.	1906	Lieutenant, Inf., N. A.
Langley, J. E.	1898	Captain, Eng.
*Layton, W. M.	1921	Corp., 57th Eng.
Lee, Arthur	1907	Captain, Inf., N. A.
LeGette, W. J.	1917	Sergeant Am. Tr., N. A.
*Lesesne, F. K.	1905	Captain, Inf., wounded.
Lindsay, J. R.	1913	Lieutenant, Inf., N. A.
Little, J. P.	1913	Signal Corps.
Littlejohn, S. C.	1914	Lieutenant, Inf., N. A.
Lumley, H.	1914	Lieutenant, Inf., N. A.
McClenaghan, G. P.	1920	Cadet West Point
McCravy, H. W.	1917	Lieutenant, F. A., N. A.
McCully, R. H.	1917	Lieutenant, Inf., N. A.
McFadden, M. S.	1919	Lieutenant, F. A., N. A.
McKittrick, S. L.	1919	Lieutenant, Inf.
*McMurran, J. P. C.	1917	Lieutenant, Marine Av. C.
*McSwain, McC.	1920	U. S. A. A.
Mackorell, H. R.	1919	Sergeant, Inf., N. A.
*Mahon, G. H.	1909	Major, Inf., N. A., wounded.
Manigault, R. S.	1914	Lieutenant, F. A., N. A.
Marshall, A.	1902	
Marshall, J. Q.	1919	Naval Mil., Chief Qm.
Mauldin, O. K.	1893	Captain, Inf.
Middleton, W. I.	1921	U. S. N.
*Mikell, F. T.	1914	Captain, Inf., wounded.
*Miler, D. S.	1917	Sergt., Am. Tr., N. A.
*Miller, J. S.	1921	Sergeant, Inf., N. A.
*Moore, F. T.	1920	Engineer, U. S. N., R. F.
Moore, M. P.	1911	Private
Moore, W. C.	1916	Captain, Inf., N. A.
Moorman, C. W.	1893	San. Tr.
Morgan, J. L.	1921	U. S. N.
*Mulloy, W. A.	1909	Lt. Inf., killed in France
Muncaster, J. H.	1906	Major, U. S. A.
Murdoch, J. H.	1010	Lieutenant, Inf., N. A.
Murray, C. S.	1914	Lieutenant, Inf., N. A.
Nichols, J. G. M.	1917	Pvt. Dental C.
O'Neal, H. M.	1921	Naval Res.
O'Neill, L. A.	1914	Ensign, U. S. N.
*Padgett, V. L.	1913	Captain, U. S. A., Reg.
*Passailaigue, E. P.	1912	Captain, Inf., N. A.
*Pearsall, H. S.	1918	Private, Motor Co.

Symbols: *, indicates service in American Expeditionary Force; R. Regular Army; G, National Guard; N, National Army; V, Navy; M, Marine Corps.

*Pearlstine, E. S.	1919	Private, M. P.
Perry, E. C.	1921	Lieutenant, Inf., N. A.
Phinney, J. H.	1911	Lieutenant, Inf., N. A.
Plowden, B. C.	1916	Captain, Inf., N. A.
Price, P. A.	1918	Lt., M. G. Co., N. A.
Quarles, R. P.	1919	Lieutenant, Inf., N. A.
Randle, E. L.	1914	Captain, F. A., N. A.
Rhame, J. M.	1912	Lieutenant, Inf., N. A.
Rice, O. G.	1917	Lieutenant, Inf., N. A.
Riley, J. M.	1912	Lieutenant, Inf., N. A.
Riley, J. W.	1904	Captain, Inf., N. A.
Rowton, C. H.	1910	Sergeant, Tr. Det.
*Salley, T. E.	1913	Lieut., Inf., N. A.
*Sease, H. S.	1913	Lieut., U. S. N.
Seibels, R. L.	1909	Captain, M. R. C. F. H.
Sherrill, C. A.	1921	Lieutenant, Inf., N. A.
Simmons, G. R.	1919	Lieutenant, U. S. A.
*Smith, E. B.	1916	Lieutenant, Inf., N. A.
Smith, H. C.	1910	Lieutenant, Inf., N. A.
Smith, W. J.	1915	Sergeant, Inf., N. A.
Snead, K. G.	1919	Lieut., Inf.
*Spigener, G. H.	1908	Captain, Inf., N. A.
*Spigener, J. V.	1917	Corporal C,. A. C., N. A.
Spratt, T. B.	1897	Lt.-Colonel, Inf., N. A.
*Springs, H. B.	1899	Colonel, U. S. N. G.
Stubbs, T. M.	1920	Lieutenant, Inf., N. A.
Taber, A. R.	1919	Lieutenant, Inf., N. A.
Tatum, J. M.	1919	Lieutenant, Inf., U. S. A.
Taylor, O. N.	1919	Lieutenant, Inf., N. A.
*Thompson, H. S.	1915	Lieutenant, Inf., N. A., wounded.
Thompson, F. E.	1920	U. S. M. C.
Vinzant, W. D.	1899	Captain, M. G. Co.
*Walker, G. W.	1915	Lieutenant, U. S. A.
*Walker, J. H.	1919	Ambulance Co.
Weeks, J. C.	1918	Lieut., Inf., N. A.
Westmoreland, W. H.	1886	Lt.-Colonel, Cav., U. S. A., Reg.
*White, R. G.	1906	Lieutenant, Inf., N. A., wounded; D. S. Cross.
Whitelaw, J. L.	1919	Lt., Inf., U. S. A., Reg.
*Wickenberg, C. H.	1916	Corp.
Williams, W. G.	1920	Lieut., Inf., N. A.
Wise, G. C.	1921	Lieut., Inf., N. A.
Witsell, P. F.	1917	Lieut., Inf., N. A.
Woodberry, J. H.	1911	Major Ord., Dept., U. S. A.
Woods, T. B.	1911	Lieut., M. R. C. A. C.
Wulburn, F. M.	1921	Litut., Inf., N. A.

#8899 ④

21528

Milton Keynes UK
Ingram Content Group UK Ltd.
UKHW012320120324
439332UK00007B/1042